SAS Press

MW00764392

The Essential Guide to
SAS® Dates and Times

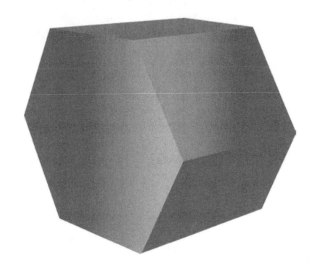

Derek P. Morgan

The correct bibliographic citation for this manual is as follows: Morgan, Derek P. 2006. *The Essential Guide to SAS® Dates and Times.* Cary, NC: SAS Institute Inc.

The Essential Guide to SAS® Dates and Times

SAS Publishing provides a complete selection of books and electronic products to help customers use SAS software to its fullest potential. For more information about our e-books, e-learning products, CDs, and hard-copy books, visit the SAS Publishing Web site at **support.sas.com/pubs** or call 1-800-727-3228.

Table of Contents

Acknowledgments

Many people helped to make this book happen. I shouldn't get all of the credit. Firstly, I'd like to give special thanks to Patsy Poole and Julie Platt from SAS Press. Their encouragement and enthusiasm kept me going. I owe a lot to Art Carpenter, SAS guru. This book would not exist without his interest, patience, and gentle direction. Another big thank you goes to Caroline Brickley, Joan Stout Knight, Candy Farrell, and Patrice Cherry, all from SAS Press, who turned my manuscript into a real book.

Thanks to Andrew Karp for introducing me to the world of PROC EXPAND, and to Mike Forno at SAS Institute's Technical Support for answering my questions on it. I greatly appreciate the American Public Transportation Association (http://www.apta.com,) who allowed me to use data that they compiled from their member transit agencies for the PROC EXPAND examples, and Erik Tilanus for sharing some of his knowledge to help improve the content of this book.

I'd also like to thank the technical reviewers from SAS Institute, Richard Bell, Chris DeHart, Rick Langston, and Kim Wilson for their thoughtful comments and corrections, as well as Michelle Schlude from SAS Institute's Technical Support for helping to iron out the more sticky points of intervals.

Closer to home, I want to acknowledge the Divisions of Biostatistics and Statistical Genomics at Washington University, especially Drs. D.C. Rao and Michael Province for their moral support, Jeanne Cashman for proofing the final draft, and Avril Adelman for giving me a SAS user's perspective.

Last, but most certainly not least, my wife Billie, and son Terec deserve a great deal of thanks for letting me spend a good number of my evenings and weekends with Microsoft Word and SAS instead of them.

CHAPTER 1

Introduction to Dates and Times in SAS

In the years that I've been working with SAS, and teaching students how to use it, I've noticed two things about it that consistently confuse programmers who are new to SAS. First, there is the "implied" DO-UNTIL (end-of-file) of the DATA step, and then there is the concept of dates (and times) within SAS. I've seen many misuses of character strings masquerading as dates and/or times over the past years. However, this is only the tip of the iceberg when it comes to the power and flexibility of dates and times in SAS. There is much more than just having numbers representing date and time values in SAS. We'll start with the basics in the first three chapters, and then progress to some more advanced uses of those date and time values.

1.1   How Does It Work? (January 1, 1960 and Midnight as Zero)

SAS has three separate counters that keep track of dates and times. The date counter started at zero on January 1, 1960. Any day before 1/1/1960 is a negative number, and any day after that is a positive number. Every day at midnight, the date counter is increased by one. The time counter runs from zero (at midnight) to 86,399.9999, when it resets to zero. The last counter is the datetime counter. This is the number of seconds since midnight, January 1, 1960. Why January 1, 1960? One story has it that the founders of SAS wanted to use the approximate birth date of the IBM 370 system, and they chose January 1, 1960 as an easy-to-remember approximation.

Many database programs maintain their dates as a value relative to some fixed point in time. This makes calculating durations easy, and working with dates stored in this fashion becomes a matter of addition, subtraction, multiplication, and division.

1.2   Internal Representation (Storage as Integers or Real Numbers)

SAS stores dates as integers, while the datetime and time counters are stored as real numbers to account for fractional seconds. The origin of the algorithm used for SAS date processing comes from a *Computerworld* article dated January 14, 1980 by Dr. Bhairav Joshi of SUNY-Geneseo. The earliest date that SAS can handle with this algorithm is January 1, 1582. The latest date is far enough into the future that four digits can't display the year.

1.3 External Representation (Basic Format Concepts)

The dates as stored by SAS don't do us much good in the real world. The statement "I was born on –242" won't mean much to anyone else. On the other hand, "May 4, 1959" can easily be translated into something that most people can understand. SAS has a built-in facility to perform automatic translation between SAS numbers and dates and times as understood by the rest of the world. This automatic translation is performed with what are called formats. Formats display the date, time, and datetime values in a fashion that is much more easily understood. Formats do *not* change the values themselves; they are just a way to display the values in any output.

What happens if you have a date or time and want to translate it into SAS date and time values? SAS has another built-in facility which performs the reverse translation, from the dates and times we understand and use to the values that SAS stores. This translation is done using informats. Informats translate what they are given into the values that are stored in SAS variables. We will discuss formats and informats in detail in Chapters 2 and 3, because there are dozens of them.

1.4 Date and Time as Numeric Constants in SAS

We've talked about internal and external representation of dates and times. How do you put a specific date into a program as a constant? Formats change the way the values are displayed in output, so you can't use them. Informats translate what they are given, so you could use them, but then you'd need to use the INPUT() function (see Section 3.3.2), which takes a value you give it and translates it with an INFORMAT. That's very inefficient. Look at the following program (Example 1.4.1) to see how date, time, and datetime constants are written into a SAS program. Take note of the quotation marks around the values for date, time, and datetime, and the letters that follow each closing quote.

The quotes are used to create a literal value. You may use a pair of single or double quotes to specify the literal value. The only difference between using single and double quotes around the date would be macro expansion. The most important part of a date constant is the letter that immediately follows the last quote. The letter "D" stands for date, "T" for time, and "DT" for datetime, and you can use either upper or lowercase. If you put a date in quotes without the letter at the end, you will create a character variable, not a numeric variable with a date, time, or datetime value. The difference might not become apparent until you try to do something with the variable you created that involves a calculation. Don't forget your "D", "T", or "DT"! Example 1.4.1 demonstrates how date constants are defined and then automatically converted to SAS date values.

Example 1.4.1 Date Constants

```
DATA date_constants;

date = '04aug2004'd;  /* This is a date constant */
time = '07:15:00't;   /* This is a time constant */
datetime = '07aug1904:21:31:00'dt;  /* This is a datetime constant */
run;

TITLE "Unformatted Constants";
PROC PRINT DATA=date_constants;
VAR date time datetime;
run;

TITLE "Formatted Constants";
PROC PRINT DATA=date_constants;
VAR date time datetime;
FORMAT date worddate32. time timeampm9. datetime datetime32.;  /* Format
the constants */
run;
```

Here is the resulting output:

Unformatted Constants		
date	time	datetime
16287	26100	-1748226540

Formatted Constants		
date	time	datetime
August 4, 2004	7:15 AM	07AUG1904:21:31:00

Without formats, you can see that the date constants we created are stored as their actual SAS date, time, and datetime values. They don't make much sense until you format them.

1.5 System Options Related to Dates

SAS has several system options; these affect the way that the SAS job or session works. There are four important options that affect dates: YEARCUTOFF, DATESTYLE, DATE/NODATE, and DTRESET.

YEARCUTOFF

On December 31, 1999, people were holding their breath. The majority of dates stored on computers allowed only two digits for the year, and assumed that the first two digits were (and would always be) "19". This didn't account for storage of dates where the first two digits of the year were not "19", and thus, the "Y2K problem" was born. How does SAS handle two-digit years? When is a two-digit year in the 1900's, and when is it in the 2000's? What if you have old data and all those dates need to be in the 1800's? What does SAS do? The answer is: *YOU* tell SAS how to handle two-digit years. There is a system option called YEARCUTOFF that lets you specify a 100-year span for two-digit years. It applies to all dates with two-digit years that you give SAS. This means that it applies to: date constants, date values read from raw data with the INPUT statement, and date values that are created from character strings with the INPUT() function. The YEARCUTOFF system option does not affect values that are stored as SAS date values, regardless of their display, so once you create a date or datetime value, YEARCUTOFF no longer has any effect on it.

The system default is 1920. This means that any two-digit year from 20 to 99 will be translated as 1920 to 1999, while years from 00 to 19 will be translated as 2000 to 2019. The syntax is:

OPTIONS YEARCUTOFF= *(y)yyyy;*   /* (y)yyyy can be from 1582 to 19900 */

Let's use a series of OPTIONS statements and date constants to illustrate. In the following program, three datasets are created with four identical date constants that use two-digit years. The only thing that changes is the value of YEARCUTOFF. Example 1.5.1 shows how YEARCUTOFF translates two-digit year values using date constants.

Example 1.5.1 How the YEARCUTOFF System Option Works

```
OPTIONS YEARCUTOFF=1920;  /* SAS System default */

DATA yearcutoff1;
date1 = "15JUL06"d;
date2 = "27FEB48"d;
date3 = "04may69"d;
date4 = "10dec95"d;
RUN;

PROC PRINT DATA=yearcutoff1;
FORMAT date1-date4 mmddyy10.;
RUN;
```

Here is the resulting output:

date1	date2	date3	date4
07/15/2006	02/27/1948	05/04/1969	12/10/1995

With the default of 1920 in effect, you can see that the first date is placed in the 21st century, while the others remain in the 20th. Let's move the 100-year period back by 80 years and see what happens.

```
OPTIONS YEARCUTOFF=1840;

DATA yearcutoff2;
date1 = "15JUL06"d;
date2 = "27FEB48"d;
date3 = "04may69"d;
date4 = "10dec95"d;
RUN;

PROC PRINT DATA=yearcutoff2;
FORMAT date1-date4 mmddyy10.;
RUN;
```

Here is the resulting output:

date1	date2	date3	date4
07/15/1906	02/27/1848	05/04/1869	12/10/1895

Now the first date is in the 20th century, and the others are in the 19th. Note that the only change to the code is in the OPTIONS statement. The value of YEARCUTOFF is 1840 instead of 1920. For the last part of this example, we'll set YEARCUTOFF to 1970, and use the same date constants with two-digit years again.

```
OPTIONS YEARCUTOFF=1970;

DATA yearcutoff3;
date1 = "15JUL06"d;
date2 = "27FEB48"d;
date3 = "04may69"d;
date4 = "10dec95"d;
RUN;

PROC PRINT DATA=yearcutoff3;
FORMAT date1-date4 mmddyy10.;
RUN;
```

Here is the resulting output:

date1	date2	date3	date4
07/15/2006	02/27/2048	05/04/2069	12/10/1995

Once again, the only difference in the code is in the OPTIONS statement. Now the 100-year range starts in 1970, which places every date except the last one in the 21st century.

As with many SAS system options, YEARCUTOFF is effective when it is encountered within the program. If you have multiple OPTIONS statements that include YEARCUTOFF= in your program, each one will affect all date constants, raw data, and date values created from character strings with the INPUT( ) function until the next OPTIONS YEARCUTOFF= statement changes the 100-year range. As an example, if you were to put the three programs in the above example together in one file, the result would be the same, as long as you did not move the OPTIONS YEARCUTOFF= statements.

DATESTYLE

This system option is important if you are using any of the following informats: ANYDTDTE., ANYDTDTM., or ANYDTTME. DATESTYLE controls how SAS will translate dates that can be interpreted in more than one way. This happens most often when you are using two-digit years.

Assuming that the OPTIONS statement specifies YEARCUTOFF=1920, does 11-01-06 mean November 1, 2006, January 6, 2011, or January 11, 2006?

DATESTYLE allows you to tell SAS how to interpret cases like this. You may specify any one of the following:

Table 1.5.1 *Values for DATESTYLE=*

MDY	Sets the default order as month, day, year. "11-01-06" would be translated as November 1, 2006	**YDM**	Sets the default order as year, day, month. "11-01-06" would be translated as June 1, 2011
MYD	Sets the default order as month, year, day. "11-01-06" would be translated as November 6, 2001	**DMY**	Sets the default order as day, month, year. "11-01-06" would be translated as January 11, 2006
YMD	Sets the default order as year, month, day. "11-01-06" would be translated as January 6, 2011	**DYM**	Sets the default order as day, year, month. "11-01-06" would be translated as June 11, 2001
LOCALE (default)	Sets the default value according to the LOCALE= system option. When the default value for the LOCALE= system option is "English_US", this sets DATESTYLE to MDY. Therefore, by default, "11-01-06" would be translated as November 1, 2006.		

DATESTYLE can be set at SAS invocation, through an OPTIONS statement, in the configuration file, or in the SAS Options window. The syntax is:

OPTIONS DATESTYLE=*order*;   */* order is one of the values from table 1.1 */*

Example 1.5.2 demonstrates the effect of the different DATESTYLE values on a given character string.

Example 1.5.2 How DATESTYLE Affects the ANYDTDTE. Informat

The following program goes through each of the possible values for DATESTYLE using the same character string 11-01-06 as input. The log shown below the program will demonstrate the differences.

```
OPTIONS DATESTYLE=mdy;
DATA _NULL_;
INPUT date anydtdte8.;
PUT "OPTIONS DATESTYLE=mdy, so date=" date mmddyy10.;
DATALINES;
11-01-06
;
RUN;

OPTIONS DATESTYLE=myd;
DATA _NULL_;
INPUT date anydtdte8.;
PUT "OPTIONS DATESTYLE=myd, so date=" date mmddyy10.;
DATALINES;
11-01-06
;
RUN;

OPTIONS DATESTYLE=ymd;
DATA _NULL_;
INPUT date anydtdte8.;
PUT "OPTIONS DATESTYLE=ymd, so date=" date mmddyy10.;
DATALINES;
11-01-06
;
RUN;

OPTIONS DATESTYLE=ydm;
DATA _NULL_;
INPUT date anydtdte8.;
PUT "OPTIONS DATESTYLE=ydm, so date=" date mmddyy10.;
DATALINES;
11-01-06
;
RUN;
```

```
OPTIONS DATESTYLE=dmy;
DATA _NULL_;
INPUT date anydtdte8.;
PUT "OPTIONS DATESTYLE=dmy, so date=" date mmddyy10.;
DATALINES;
11-01-06
;
RUN;

OPTIONS DATESTYLE=dym;
DATA _NULL_;
INPUT date anydtdte8.;
PUT "OPTIONS DATESTYLE=dym, so date=" date mmddyy10.;
DATALINES;
11-01-06
;
RUN;

OPTIONS DATESTYLE=locale; /* LOCALE=EN_US */
DATA _NULL_;
INPUT date anydtdte8.;
PUT "OPTIONS DATESTYLE=locale, so date=" date mmddyy10.;
DATALINES;
11-01-06
;
RUN;
```

The Log

```
OPTIONS DATESTYLE=mdy, so date=11/01/2006
OPTIONS DATESTYLE=myd, so date=11/06/2001
OPTIONS DATESTYLE=ymd, so date=01/06/2011
OPTIONS DATESTYLE=ydm, so date=06/01/2011
OPTIONS DATESTYLE=dmy, so date=01/11/2006
OPTIONS DATESTYLE=dym, so date=06/11/2001
OPTIONS DATESTYLE=locale, so date=11/01/2006
```

As you can see, DATESTYLE can have an enormous effect when the ANYDTDTE. (or ANYDTDTM. or ANYDTTM.) informats are used.

DATE/NODATE

By default, the DATE system option is in effect when you start SAS, which causes the date and time that the SAS job (or session) started to appear on each page of the SAS log and SAS output. These values are obtained from the operating system clock. If you are running SAS interactively, then, the date and time are printed only on the output, not the log. If you don't want the date and time to appear, use the NODATE system option. The syntax is:

OPTIONS NODATE;

If you've turned off DATE, then you can turn it back on with:

OPTIONS DATE;

Example 1.5.3 shows what happens to the title line printed by SAS when you use DATE and NODATE. Remember that, by default, DATE is in effect when you start SAS.

Example 1.5.3 DATE/NODATE

This is a sample of a title line with the DATE system option:

```
The SAS System            17:20 Thursday, August 5, 2004   1
```

This is what NODATE does to that title line:

```
The SAS System                                             1
```

DTRESET

If the DATE option is enabled, SAS prints the date and time that the current SAS session started. If you want a more accurate date and time on those pages, you can use the DTRESET system option. This will cause SAS to get the date and time from the operating system clock each time a page is written. That date and time will then be put on the page instead of the time that the SAS job started. Since the time is displayed in hours and minutes, you will see it change each minute only. The syntax is:

OPTIONS DTRESET;

1.6   Length and Numeric Requirements for Date, Time, and Datetime

Since dates are stored as integers, you can take advantage of that to save space when you create variables to store them. Instead of using the default length of 8 for numeric variables, set the LENGTH of the numeric variables where you are storing the dates to 4. This will safely store dates from January 1, 1582 (the earliest date SAS can handle), to October 23, 7701. A length of 5 is overkill, although that would extend the ending date another 534,773,760 days! A length of 3 will not accurately store dates outside the range of January 1, 1960 and September 13, 1960. If you declare your date variables to be a length of 4, you will be able to store two dates in the space it would take to store one if you were using the SAS default length for numeric variables.

Times may present a little bit of a problem, since times have the possibility of having decimal parts. You can get away with storing times in the same magic length of 4 that you can use for dates, and the rule is simple enough: if you want fractional seconds in your time values, use a length of 8 for maximum precision. Otherwise, the same length of 4 will store every possible whole second from midnight to midnight.

Datetime values need to be a little longer; a length of 4 will not store a datetime value with accuracy, regardless of whether you want decimal places. The number is just too big. Use a length of 6 to store datetime values; this will accurately represent datetime values (without fractions of seconds) from midnight, January 1, 1582 to 3:04:31 PM on April 9, 6315. Note that a length of 6 might not translate into other databases.

CAUTION

In all the above cases, the minimum lengths for accuracy have been given to you; do not attempt to save more space by shrinking the variables further. You will lose precision, and this could lead to unexpected results. Example 1.6.1 shows what can happen if you do not use enough bytes to store your date values.

Example 1.6.1 The Effect of LENGTH Statements on Dates

```
DATA date_length;
LENGTH len3 3 len4 4 len5 5;
len3 = '05AUG2004'd+1;
len4 = '05AUG2004'd+1;
len5 = '05AUG2004'd+1;
FORMAT len3 len4 len5 mmddyy10.;
RUN;

PROC PRINT DATA=date_length;
RUN;
```

Here is the resulting output. Notice that the date in len3 is different from the one in the other two variables. This is what can happen when you shrink the size of the variable too much. Instead of August 6, 2004, the value is wrong.

len3	len4	len5
08/05/2004	08/06/2004	08/06/2004

CHAPTER 2

Displaying SAS Date, Time, and Datetime Values as Dates and Times as We Know Them

SUMMARY

SAS date, time, and datetime values are stored as integers (unless you are storing fractional parts of seconds). They are all counted from a fixed reference point. SAS date values increment by 1 at midnight of each day, while SAS datetime values increment by 1 every second. SAS time values start at zero at midnight of each day, and increment by 1 each second.

This scheme makes it easy to calculate durations in days and seconds, but it does not do much for figuring out what a given SAS date, time, or datetime value means in terms of how we talk about them. Therefore, SAS provides a facility that makes it easy to perform the translation from SAS into the common terminology of months, days, years, hours, and seconds. The translation is done through formats.

Formats are what SAS uses to control the way data values are displayed. They can also be used to group data values together for analysis. They are essential to dates and times in SAS because SAS does not store dates and times in an easily recognizable form, as we discussed in Chapter 1. SAS has many built-in formats to display dates, times, and datetime values. Here's your handy guide to all of the date, time, and datetime formats readily available in SAS. In addition, if any of these built-in formats don't fit your needs, you have the ability to create (and store for future use) your own formats, which is covered in Sections 2.5 and 2.6.

If you are looking for a quick reference, you can go to Appendix A, which lists all of the date, time, and datetime formats, and gives a sample display using their default lengths. If the default does not give you what you want, Section 2.4 discusses each date, time, and datetime format in detail, including how to specify the length of the format, and how that length affects the display.

2.1 How Do I Use a Format?

Formats are easy to use. You can permanently associate a format with a variable by using a FORMAT statement in a DATA step as shown in Example 2.1.1.

Example 2.1.1 Permanently Associating a Format with a Variable

```
DATA test;
LENGTH date1 time1 4;
date2 = 16048;
time2 = 733000;
FORMAT date1 MMDDYY10. time1. TIMEAMPM11.;
RUN;
```

This example will create a dataset called TEST, which has two variables, date1, and time1. By using the FORMAT statement here, you have specified that whenever the values from this dataset are displayed, the values stored in the variable date1 will always be displayed with the format MMDDYY10., and those stored in time1 will always be displayed using the TIMEAMPM11. format.

If you don't want to have your data values permanently associated with a format, then you can just apply the format when you are actually writing the values to your output. The same FORMAT statement is used, but the location has changed, from the DATA step to the PROC step. Example 2.1.2 illustrates this.

Example 2.1.2 Associating a Format with a Variable for the Duration of a Procedure

```
DATA test2;
LENGTH date2 time2 4;
date2 = 16048;
time2 = 733000;
RUN;

PROC PRINT DATA=test2;
FORMAT date2 DATE9. time2 TIMEAMPM11.;
RUN;
```

Now, although there is no format assigned to either date2 or time2 in the DATA step, you have told the PRINT procedure to write these values using the two formats listed. There's another handy thing about using the FORMAT statement with a SAS procedure: if you use the FORMAT statement in a SAS procedure, it will override any format that has been permanently associated with the variables for the duration of that procedure. To illustrate, we'll take the dataset "test" from Example 2.1.2 above. The variables date1 and time1 have been associated with the formats MMDDYY10. and TIMEAMPM11., respectively. What if your report needs the date printed out with the day of the week, month name, day, and year, while the time needs to be seconds after midnight? The PROC PRINT step will look like this:

```
PROC PRINT DATA=test;
FORMAT date1 WEEKDATE37. time1;
RUN;
```

All SAS procedures will use the formats specified in the FORMAT statement that is part of the PROC step instead of the formats associated with the variable in the dataset. Therefore, in the above example, date1 will be printed with the WEEKDATE. format. What about time1? There's no format name given after the variable name in the FORMAT statement. This is how to tell SAS not to use any formats that may be associated with the variable. To remove a FORMAT from a variable, make sure that no format names of any kind follow it anywhere in the FORMAT statement. In the following code segment, both time1 and date1 will be formatted with WEEKDATE37.

```
PROC PRINT DATA=test;
FORMAT time1 date1 WEEKDATE37.;
RUN;
```

2.2 So Just How Many Built-in Formats Are There for Dates and Times?

The answer is lots. We will discuss each of them in detail here, but if you're looking for a quick reference, see Appendix A. SAS formats have their own syntax structure. There is a format name, followed by a width specification, and they all end with a period. The period is critical – it is what allows SAS to recognize the word as a format, and not some other SAS keyword, or text. The width specification varies with each format. This is very important to dates, because SAS will make abbreviations to the displayed value if you do not specify enough characters for the width specification. The abbreviation that SAS will use may not give you the output that you want. Each format has its own default width, which is what SAS will use if you do not specify a width. The default width is noted in the description for each format below, and it is usually the width that will accommodate the longest value to be displayed. For example, the default width for the DOWNAME. format is 9. That will accommodate the string Wednesday, which is the longest English day-of-week name.

2.3 A Quick Note About Date Formats, Justification, and ODS

Each date format has a default justification with respect to the width specification that you give it. Since numeric values are right-justified in SAS, most formats that are applied to date, time or datetime formats are right-justified, with a few exceptions. This is only applicable to traditional column-based output. In ODS destinations other than LISTING, values are justified within a table column by the SAS procedure default or by a user-defined ODS template. By default, SAS makes its columns wide enough to fit the widest item in a given column. Therefore, any leading spaces caused by specifying a width that is too wide to fit the formatted value won't show up in ODS output.

If you do not specify column alignment in an ODS template, certain ODS destinations (such as RTF and PDF) will justify values within a column according to the justification of the format used in the column, without leading spaces.

2.4 Detailed Discussion of Each Format

This section will give a detailed explanation of all the current standard formats available for SAS date, time, and datetime values. In addition to the display that results from using a given format, the explanation includes information on the default width specification and its possible values, annotated examples of the display with varying width specifications, and usage notes. Date formats will be covered first, then time and datetime formats. Each subsection is arranged alphabetically.

2.4.1 Date Formats

A date format translates SAS date values into one of several easily recognized equivalents. You may specify the width (number of characters) that the translated text will occupy, but each format has its own default width specification, shown as **w** in this text. The default width specification is given in the description of each format. Some, but not all, of the date formats allow you to specify the character that separates each element of the date. You must not use a date format to translate datetime values. If you try to translate a datetime value with a date format, you will get incorrect output. (For an example, see Section 2.4.3.)

DATE*w*.   Writes dates as the numerical day of the month followed by the three-letter month abbreviation and the year, without any separating characters. It is right-justified within the field. ***w*** can be from 5–9, and the default width is 7. If you want to display four-digit years, use DATE9. The following table shows the result when the date value is 8449, which corresponds to February 18, 1983. Note the alignment of the date printed.

Format Name	Result	Comment
DATE.	18FEB83	
DATE5.	18FEB	No room for year to be displayed.
DATE6.	18FEB	Output is moved to the right by 1 space.
DATE7.	18FEB83	
DATE8.	18FEB83	Two-digit year, leading space.
DATE9.	18FEB1983	

This format is analogous to the DTDATE. format, which displays datetime values in the same manner.

DAY*w*.   Writes the numerical day of the month, and it is right-justified within the field. ***w*** can be from 2–32, and the default width is 2. Specifying anything longer than 2 will only place more spaces in the field to the left of the number, so it's not necessary to specify more than 2. The following table shows the result when the date value is 16739, which corresponds to October 30, 2005:

Format Name	Result	Comment
DATE2.	30	
DATE5.	30	See how the day is moved three spaces to the right.

DDMMYY*w*.   Writes dates as day/numerical month/year, where the slash (/) is the separator, and it is right justified within the field. **w** can be from 2–10, and the default width is 8. If you specify a width from 2–5, the date will be truncated on the right, with SAS trying to fit as much of the day and month as possible in the space allowed. If you use 6, no slashes will be printed. A width of 7 will cause SAS to print a two-digit year without a slash, and widths of 8 or 9 will put a two-digit year after the slashes. Use 10 to get a four-digit year with slashes. The following table shows the result when the date value is 15486, which corresponds to May 26, 2002:

Format Name	Result	Comment
DDMMYY5.	26/05	
DDMMYY6.	260502	
DDMMYY7.	260502	Moved right 1 space.
DDMMYY8.	26/05/02	
DDMMYY9.	26/05/02	Still a two-digit year, moved right 1 space.
DDMMYY10.	26/05/2002	

DDMMYY*xw*.   Is similar to DDMMYY*w*., above. It is also right-justified. However, with this format, you can specify what character separates the day, numerical month, and year. The **x** in the format name represents the separator between the day, month, and year. **x** can be:

x	Character Displayed in Output	Comment
B	blank	
C	colon (:)	
D	dash (—)	
N	no separator	**w** is a maximum of 8, not 10.
P	period (.)	
S	slash (/)	Effectively the same as using the DDMMYY*w*. format.

w can be from 2–10, with the default being 8. This works the same way as the DDMMYY. format with respect to what SAS fits in the space specified. Again, if you specify a width from 2–5, the date will be truncated on the right, with SAS trying to fit as much of the day and month as possible in the space allowed. If you use 6, no separator will be used. At 7, SAS will print a two-digit year without separator, and 8 or 9 will put a two-digit year after the separator. Use 10 to get a four-digit year with your separator. The following table shows the result when the date value is 16110, which corresponds to February 9, 2004:

Format Name	Result	Comment
DDMMYYP5.	09.02	Only space for day and month
DDMMYYB6.	090204	Not enough space for the blank separator.
DDMMYYN7.	090204	No separator.
DDMMYYD8.	09-02-04	Enough space for two-digit year.
DDMMYYS9.	09/02/04	Enough space for two-digit year, same as using DDMMYY9. format.
DDMMYYC10.	09:02:2004	

DOWNAMEw. Writes the date as the name of the day of the week. It is right-justified, so if you give it too much space, there will be leading blanks. **w** can be from 1 to 32, and the default is 9. If you don't specify **w**, SAS will print the entire name of the day. On the other hand, if you don't have enough space, SAS will truncate the name of the day to fit. The following table shows the result when the date value is 17361, which corresponds to Saturday, July 14, 2007:

Format Name	Result	Comment
DOWNAME5.	Satur	
DOWNAME6.	Saturd	
DOWNAME7.	Saturda	
DOWNAME8.	Saturday	
DOWNAME9.	Saturday	"Saturday" is only 8 characters long, so 1 leading blank is added, making it appear as if it has been moved to the right.
DOWNAME10.	Saturday	2 leading blanks added.
DOWNAME11.	Saturday	3 leading blanks added.

JULDAYw. Writes the date as the Julian day of the year, which is a value from 1 to 366. It is right-justified. **w** can be from 3 to 32, and the default is 3.

Format Name	Result	Comment
JULDAY3.	91	There is a leading space here because there are fewer than 3 digits in the value displayed. The date value used here is 17988, which corresponds to April 1, 2009.
JULDAY3.	107	There is no leading space here because there are 3 digits in the value displayed. The date value used here is 18004, which corresponds to April 17, 2009.
JULDAY4.	91	2 leading spaces. You will also have 2 leading spaces there are fewer than 2 digits and you specify **w** as 3. The date value used here is 17988, which corresponds to April 1, 2009.
JULDAY.	91	Leading space. Default is 3. The date value used here is 17988, which corresponds to April 1, 2009.

JULIANw. Writes your date value as a Julian date, with the year preceding the Julian day. It is right-justified. **w** can be from 5 to 7, and the default is 5. If you specify a width of 5, the year portion of the Julian date is two digits long, while specifying 6 will give you one leading blank and a two-digit year. If you specify a width of 7, the year portion is four digits long. The following table shows the result when the date value is 18004, which corresponds to April 17, 2009:

Format Name	Result	Comment
JULIAN5.	09107	
JULIAN6.	09107	One leading blank added.
JULIAN7.	2009107	

MMDDYYw. Writes the date as numerical month/day/year, where a slash (/) is the separator. It is right-justified within the field. **w** can be from 2 to 10, and the default is 8. It is similar to the DDMMYY. format in that if you specify 2–5 for the width, the date will be truncated on the right, with SAS trying to fit as much of the day and month as possible in the space allowed. If you use 6, no slashes will be printed, but it will print a two-digit year. At 7, SAS will print a two-digit year without a slash, and 8 or 9 will put a two-digit year after the slashes. Use a width of 10 to get a four-digit year with slashes. The following table shows the result when the date value is 16773, which corresponds to December 3, 2005:

Format Name	Result	Comment
MMDDYY2.	12	Month only.
MMDDYY3.	12	Month, with 1 leading space.
MMDDYY4.	1203	Month and day, no separator.
MMDDYY5.	12/03	Month and day with separating slash.
MMDDYY6.	120305	Month, day, and year, no separator.
MMDDYY7.	120305	Month, day, and year, no separator, leading blank.
MMDDYY8.	12/03/05	
MMDDYY9.	12/03/05	Still a two-digit year, leading blank added.
MMDDYY10.	12/03/2005	

MMDDYYxw. Displays the date in the same way that the MMDDYY. format does, except that you can specify the separator. The **x** in the format name specifies the separator that you want to use according to the following table:

x	Character Displayed in Output	Comment
B	blank	
C	colon (:)	
D	dash (—)	
N	no separator	**w** is a maximum of 8, not 10.
P	period (.)	
S	slash (/)	Effectively the same as using the MMDDYY*w*. format.

The date will be right-justified within the width you specify. **w** can be from 2 to 10, and the default is 8. If you specify 2–5, the date will be truncated on the right, with SAS trying to fit as much of the day and month as possible in the space allowed. If you use a width of 6, no separator will be used. At 7, SAS will print a two-digit year without a separator, and widths of 8 or 9 will put a two-digit year after the separator. Use 10 to get a four-digit year with separators. The following table shows the result when the date value is 15997, which corresponds to October 19, 2003:

Format Name	Result	Comment
MMDDYYD5.	10-19	No room for year.
MMDDYYS6.	101903	No room for separators.
MMDDYYN7.	101903	Leading space, no separator specified, not enough room for four-digit year.
MMDDYYC8.	10:19:03	Colon as separator, still two-digit year.
MMDDYYP9.	10.19.03	Not enough room for four-digit year, leading space added.
MMDDYYB10.	10 19 2003	

MMYYw. Displays the zero-filled month number and year for the given date value, separated by the letter M. It is right-justified, and **w** can be from 5 to 32, with a default width of 7. When **w** is specified as 5 or 6, a two-digit year is used. If **w** is 7 or more, a full four-digit year is displayed. Since this format can only display a maximum of 7 characters, a width greater than 7 will just add leading spaces. The following table shows the result when the date value is 16834, which corresponds to February 2, 2006:

Format Name	Result	Comment
MMYY5.	02M06	
MMYY6.	02M06	Two-digit year, leading space.
MMYY7.	02M2006	Four-digit year.
MMYY8.	02M2006	Four-digit year, 1 leading space.
MMYY9.	02M2006	Four-digit year, 2 leading spaces.

MMYYxw. Displays the month number and year for a given date value in the same fashion as the MMYY format, except that you may specify the separator with **x**, according to the table below. Note that the blank is not valid with this format, while it is valid with the **DDMMYYxw.** and **MMDDYYxw.** formats.

x	Character Displayed in Output	Comment
C	colon (:)	
D	dash (—)	
N	no separator	**w** can be from 4–32, with a default of 6.
P	period (.)	
S	slash (/)	

It is right-justified, and **w** can be from 5 to 32, with a default width of 7. When **w** is specified as 5 or 6, a two-digit year is used. If **w** is 7 or more, a full four-digit year is displayed. Specifying no separator with N will change the range of **w** from 4 to 32, and the default is changed to 6. Since this format can only display a maximum of 7 characters, anything more than 7 will just add leading spaces. The following table shows the result when the date value is 15975, which corresponds to September 27, 2003:

Format Name	Result	Comment
MMYYN5.	0903	1 leading space, no separator, two-digit year.
MMYYS6.	09/03	1 leading space, two-digit year.
MMYYD7.	09-2003	Four-digit year.
MMYYC8.	09:2003	Four-digit year, 1 leading space.
MMYYP9.	09.2003	Four-digit year, 2 leading spaces.

MONNAMEw. Displays the name of the month. It is right-justified, and **w** can be from 1 to 32, with a default of 9. Using a value greater than 9 will only add leading spaces. SAS will truncate the month name as necessary to fit in the width. The following table shows the result when the date value is 16338, which corresponds to September 24, 2004:

Format Name	Result	Comment
MONNAME3.	Sep	Specifying a **w** of 3 will display the 3-letter month abbreviation.
MONNAME4.	Sept	
MONNAME5.	Septe	
MONNAME6.	Septem	
MONNAME7.	Septemb	
MONNAME8.	Septembe	
MONNAME9.	September	
MONNAME10.	September	Leading space.

MONTHw. Displays the number of the month of the year. It is right-justified, and **w** can be from 1 to 21, with a default of 2. Using a **w** of 1 will display the month number as a hexadecimal value (1 through C). The following table shows the result when the date value is 16773, which corresponds to December 3, 2005:

Format Name	Result	Comment
MONTH1.	C	**w** of 1 always prints a single hexadecimal digit.
MONTH2.	12	
MONTH3.	12	1 leading space.
MONTH4.	12	2 leading spaces.

MONYYw. Displays the three-letter month abbreviation, followed by the year without any separating characters. It is right-justified, and **w** can be from 5 to 7, with a default of 5. Specifying a width of 5 will give you a two-digit year. A width of 6 will give you a two-digit year and one leading space in the displayed date, while 7 will give you a four-digit year. It is analogous to the DTMONYY format, which is used with datetime values. The following table shows the result when the date value is 15323, which corresponds to December 14, 2001:

Format Name	Result	Comment
MONYY5.	DEC01	
MONYY6.	DEC01	Two-digit year, 1 leading space.
MONYY7.	DEC2001	

PDJULGw. Writes a packed Julian date in hexadecimal format for IBM computers. Justification is not an issue, and **w** can range from 3 to 16. The default width is 4. The Julian date is written as follows: the four-digit Gregorian year is written in the first two bytes, and the three-digit integer that represents the day of the year is in the next one-and-a-half bytes. The last half-byte contains all binary 1's, which indicates the value is positive.

If the SAS date value being translated by this format is a date constant with a two-digit year, it will be affected by the YEARCUTOFF option. Look at the following SAS log.

The Log

```
OPTIONS YEARCUTOFF=1880;
 data _null_;
 date1 = "15JUN2004"d;
 date2 = "15JUN04"d;    /* Affected by YEARCUTOFF option */
 juldate1 = PUT(date1,pdjulg4.);
 juldate2 = PUT(date2,pdjulg4.);
 PUT juldate1= $hex8.;
 PUT juldate2= $hex8.;
 run;


 juldate1=2004167F
 juldate2=1904167F
```

PDJULIw.   Writes a packed Julian date in hexadecimal format for IBM computers. It only differs from the PDJULG. format in that it writes the century in the first byte as a two-digit integer, followed by two digits of the year in the second byte. The next one-and-a-half-bytes store the three-digit integer that corresponds to the day of the year, while the last half byte is filled with hexadecimal 1's that indicate a positive number. As with the PDJULG. format, justification is not an issue, and the default width is 4, with a width range of 3 to 16.

The century and year are calculated by subtracting 1900 from the four-digit Gregorian year. A year value of 1980 gives a century/year value of 0080 (1980-1900=80), while 2015 gives 0115 (2015-1900=115). Be aware that this format will *not* produce correct results for years preceding 1900. The example below demonstrates.

```
OPTIONS YEARCUTOFF=2000;
DATA _NULL_;
date1 = "15JUN1804"d;
date2 = "15JUN1996"d;
date3 = "15JUN96"d;
juldate1 = PUT(date1,pdjuli4.);
juldate2 = PUT(date2,pdjuli4.);
juldate3 = PUT(date3,pdjuli4.);
PUT juldate1= $hex8.;
PUT juldate2= $hex8.;
PUT juldate3= $hex8.;
should_be_date1 = INPUT(juldate1,pdjuli4.);
```

```
PUT should_be_date1= mmddyy10.;
PUT should_be_date1= ;
RUN;
```

Here is the resulting output:

```
juldate1=009609DF

juldate2=0096167F

juldate3=0196167F

should_be_date1=04/12/1996

should_be_date1=13251
```

Date1 is in 1804, causing the PDJULI. representation of date1 (juldate1) to be incorrect. While it should be the same day of the year (167) as date2 and date3, you can see that the day is incorrectly written as 09D, while the century value is also incorrect, marked as 00 when, by the algorithm, it should be expressed as a negative number. This is verified by using the PDJULI. informat to read juldate1, which gives a result of April 12, 1996, when it should be June 15, 1804. (This is because there is no sign bit for julday1 to indicate that the value should be negative.) The difference between date2 and date3 is caused by the YEARCUTOFF option. The two-digit year 96 is translated as 2096, not 1996 because of the option.

QTRw. Writes a date value as the quarter of the year. It is right-justified, and **w** can range from 1 to 32, with a default of 1. Since this format will only write 1 character, specifying a width greater than 1 will just add leading spaces. The following table shows the result when the date value is 12785, which corresponds to January 2, 1995:

Format Name	Result	Comment
QTR1.	1	
QTR3.	1	Two leading spaces.
QTR6.	1	Five leading spaces.

QTRR*w*. Also writes a date value as the quarter of the year, except that it displays the quarter as a Roman numeral. It is right-justified, and *w* can range from 3 to 32, with a default of 3. This format will write a maximum of 3 characters. A width specification greater than 3 will add leading spaces, as shown here:

Format Name	Result	Comment
QTRR3.	III	The date value used is 14139 (September 16, 1998.)
QTRR3.	IV	With a date value of 14229 (December 16, 1998), there is 1 leading space.

WEEKDATE*w*. Writes date values as day-of-week name, month name, day, and year. It is right-justified, and *w* can range from 3 to 37. The default is 29, which is the maximum width of a date in this format. Specifying anything longer than 29 will cause leading spaces to be added. If the width specified is too small to display the complete day of the year and month, SAS will abbreviate. The following table shows the result when the date value is 15972, which corresponds to September 24, 2003:

Format Name	Result	Comment
WEEKDATE3.	Wed	Three letter day-of-week abbreviation.
WEEKDATE5.	Wed	Two leading spaces.
WEEKDATE9.	Wednesday	Will fit all day of week names. Leading spaces will be added for days other than "Wednesday".
WEEKDATE17.	Wed, Sep 24, 2003	Full date information, abbreviated.
WEEKDATE20.	Wed, Sep 24, 2003	Three leading spaces, date abbreviated.
WEEKDATE23.	Wednesday, Sep 24, 2003	Full day-of-week name, month name abbreviated.
WEEKDATE29.	Wednesday, September 24, 2003	
WEEKDATE30.	Wednesday, September 24, 2003	Leading space.

WEEKDATX*w*. Writes date values as day-of-week name, day, month name, and year. It differs from the WEEKDATE. format in that the day of the month precedes the month name. It is right-justified, and **w** can range from 3 to 37. The default is 29, which is the maximum width of a date in this format. Specifying anything longer than 29 will cause leading spaces to be added. If the width specified is too small to display the complete day of the year and month, SAS will abbreviate. The following table shows the result when the date value is 15972, which corresponds to September 24, 2003:

Format Name	Result	Comment
WEEKDATX3.	Wed	Three letter day-of-week abbreviation.
WEEKDATX5.	Wed	Two leading spaces.
WEEKDATX9.	Wednesday	Will fit all day of week names. Leading spaces will be added for days other than "Wednesday".
WEEKDATX17.	Wed, 24 Sep, 2003	Full date information, abbreviated.
WEEKDATX20.	Wed, 24 Sep, 2003	Three leading spaces, date abbreviated.
WEEKDATX23.	Wednesday, 24 Sep, 2003	Full day-of-week name, month name abbreviated.
WEEKDATX29.	Wednesday, 24 September, 2003	
WEEKDATX30.	Wednesday, 24 September, 2003	Leading space.

WEEKDAY*w*. Writes the date value as the number of the day of the week, where 1=Sunday; 2=Monday, etc.). It is right-justified, and **w** can be from 1 to 32. The default is 1. Since the maximum width of the display is always one character, specifying anything more will just cause leading spaces to be added. The following table shows the result when the date value is 12533, which corresponds to Monday, April 25, 1994:

Format Name	Result	Comment
WEEKDAY1.	2	
WEEKDAY2.	2	One leading space.
WEEKDAY3.	2	Two leading spaces.

WEEKU*w*. Writes the date value as a week number in decimal format using the U algorithm. Unlike many other date, time, and datetime formats, it is *left-justified*. *w* can be from 1 to 200, and the default is 11. Specifying any value greater than 11 will display the same results as if *w* were 11. The U algorithm calculates weeks based on Sunday being the first day of the week, and the week number is displayed as a two-digit number from 0 to 53, with a leading zero if necessary. The display that this format presents varies, based on the width specification. The following table shows the result when the date value is 17232, which corresponds to March 7, 2007, which was a Wednesday in the ninth week of the year:

Format Name	Result	Comment
WEEKU3.	W09	"W" indicates week, week number follows, leading zero if necessary.
WEEKU4.	W09	Same as WEEKU3. No leading spaces.
WEEKU5.	07W09	Two-digit year precedes week.
WEEKU6.	07W09	Same as WEEKU5.
WEEKU7.	07W0904	Two-digit year precedes week, week followed by the number of the day of the week.
WEEKU8.	07W0904	Same as WEEKU7.
WEEKU9.	2007W0904	Four-digit year precedes week, week number is followed by number of the day of the week.
WEEKU10.	2007W0904	Same as WEEKU9.
WEEKU11.	2007-W09-04	Separator added between year, week number, and number of the day of the week.
WEEKU12.	2007-W09-04	Same as WEEKU11.

WEEKVw. Writes the date value as a week number in decimal format using the V algorithm, which is International Standards Organization (ISO) compliant. It is *left-justified* in the same fashion as the WEEKU. format. **w** can be from 1 to 200, and the default is 11. Specifying any value greater than 11 will display the same results as if **w** were 11. The V algorithm calculates weeks based on Monday being the first day of the week, and the week number is displayed as a two-digit number from 0 to 53, with a leading zero if necessary. In addition, the first week of the year contains both January 4 and the first Thursday of the year. Therefore, if the first Monday of the year falls on January 2, 3, or 4, the preceding days of the calendar year are considered a part of week 53 of the previous calendar year. The following table shows the result when the date value is 15340, which corresponds to December 31, 2001. Note that although the date is in 2001, the algorithm used by this format places the date in the year 2002. Monday, December 31, 2001, is considered to be the first day of the first week of the year 2002.

Format Name	Result	Comment
WEEKV3.	W01	"W" indicates week, week number follows, leading zero if necessary.
WEEKV4.	W01	Same as WEEKV3. No leading spaces.
WEEKV5.	02W01	Two-digit year precedes week.
WEEKV6.	02W01	Same as WEEKV5.
WEEKV7.	02W0101	Two-digit year precedes week, week followed by the number of the day of the week.
WEEKV8.	02W0101	Same as WEEKV7.
WEEKV9.	2002W0101	Four-digit year precedes week, week number is followed by number of the day of the week.
WEEKV10.	2002W0101	Same as WEEKV9.
WEEKV11.	2002-W01-01	Separator added between year, week number, and number of the day of the week.
WEEKV12.	2002-W01-01	Same as WEEKV11.

WEEKWw.   Writes the date value as a week number in decimal format using the W algorithm. As with the WEEKU. and WEEKV. formats, it is *left-justified*. **w** can be from 1 to 200, and the default is 11. Specifying any value greater than 11 will display the same results as if **w** were 11. The W algorithm calculates weeks based on Monday being the first day of the week without any other restriction. The week number is displayed as a two-digit number from 0 to 53, with a leading zero if necessary. The display that this format presents varies, based on the width specification. The following table shows the result when the date value is 15340, which corresponds to December 31, 2001, (the same date used in the V algorithm example above). Note that the W algorithm assigns the date as the first day of the last week of the calendar year 2001.

Format Name	Result	Comment
WEEKW3.	W53	"W" indicates week, week number follows, leading zero if necessary.
WEEKW4.	W53	Same as WEEKW3. No leading spaces.
WEEKW5.	01W53	Two-digit year precedes week.
WEEKW6.	01W53	Same as WEEKW5.
WEEKW7.	01W5301	Two-digit year precedes week, week followed by number of day of week.
WEEKW8.	01W5301	Same as WEEKW7.
WEEKW9.	2001W5301	Four-digit year precedes week, week number is followed by number of the day of the week.
WEEKW10.	2001W5301	Same as WEEKW9.
WEEKW11.	2001-W53-01	Separator added between year, week number, and number of the day of the week.
WEEKW12.	2001-W53-01	Same as WEEKW11.

WORDDATEw.   Displays the date value as name-of-month, day, and year. It is right-justified, and **w** can range from 3 to 32. The default is 18. If the width specified is less than 18, SAS will abbreviate and add leading spaces as necessary, regardless of whether the specific date to be displayed will fit in the allocated space because of its value. The following table shows the result when the date value is 15945, which corresponds to August 28, 2003:

Format Name	Result	Comment
WORDDATE3.	Aug	
WORDDATE12.	Aug 28, 2003	
WORDDATE15.	Aug 28, 2003	Three leading spaces.
WORDDATE18.	August 28, 2003	Full month-name, but only 15 characters, therefore, 3 leading spaces.
WORDDATE20.	August 28, 2003	Five leading spaces.

WORDDATX*w*. Displays the date value as day, name-of-month, and year. It differs from the WORDDATE. format in that the day precedes the name-of-month. It is right-justified, and **w** can range from 3 to 32. The default is 18. If the width specified is less than 18, SAS will abbreviate and add leading spaces as necessary, even if the date to be displayed will fit in the width specified. In the following table, you see that March is abbreviated for width specifications less than 18, even though there is room to print the entire date string. The table shows the result when the date value is 16144, which corresponds to March 14, 2004:

Format Name	Result	Comment
WORDDATX3.	Mar	
WORDDATX12.	14 Mar 2004	Leading space.
WORDDATX14.	14 Mar 2004	**w** is less than 18, so the format uses the abbreviated month name, and adds leading spaces even though the text displayed would fit in 13 characters.
WORDDATX16.	14 Mar 2004	**w** is still less than 18, so "March" is still abbreviated, and more leading spaces are added.
WORDDATX18.	14 March 2004	Printed with leading spaces because date string is only 13 characters long.

YEARw. Displays the year for the given date value. It is right-justified, and **w** can be from 2 to 4, with a default width of 4. When **w** is specified as 2 or 3, a two-digit year is used. The following table shows the result when the date value is 18599, which corresponds to December 3, 2010:

Format Name	Result	Comment
YEAR2.	10	Two-digit year.
YEAR3.	10	Two-digit year with a leading space.
YEAR4.	2010	Four-digit year.

YYMMw. Displays the year and month number for the given date value, separated by the letter M. It is right-justified, and **w** can be from 5 to 32, with a default width of 7. When **w** is specified as 5 or 6, a two-digit year is used. If **w** is 7 or more, a full four-digit year is displayed. Since this format can only display a maximum of 7 characters, anything more than 7 will just add leading spaces. The following table shows the result when the date value is 14476, which corresponds to August 20, 1999:

Format Name	Result	Comment
YYMM5.	99M08	Two-digit year.
YYMM6.	99M08	Leading space.
YYMM7.	1999M08	Four-digit year.
YYMM8.	1999M08	Leading space.

YYMMxw. Displays the year and month number for a given date value in the same manner as the YYMM. format above, except that you may specify the separator with **x** according to the table shown here. Unlike the **DDMMYYxw.**, **MMDDYYxw.**, and the **YYMMDDxw.** formats, a blank is not a valid separator with this format.

x	Character Displayed in Output	Comment
C	colon (:)	
D	dash (—)	
N	no separator	**w** can be from 4–32, with a default of 6.
P	period (.)	
S	slash (/)	

It is right-justified, and **w** can be from 5 to 32, with a default width of 7. When **w** is specified as 5 or 6, a two-digit year is used. If **w** is 7 or more, a full four-digit year is displayed. Specifying no separator with "N" will change the range of **w** from 4 to 32, and the default width becomes 6. Since this format can only display a maximum of 7 characters, anything more than 7 will just add leading spaces. The following table shows the result when the date value is 16315, which corresponds to September 1, 2004:

Format Name	Result	Comment
YYMMN4.	0409	No separator, minimum width is 4, two-digit year.
YYMMC5.	04:09	
YYMMD6.	04-09	One leading space, two-digit year.
YYMMP7.	2004.09	Four-digit year.
YYMMS8.	2004/09	Four-digit year, 1 leading space.

YYMMDDw. This format is a variation on the DDMMYY. and MMDDYY. formats. It writes the date as year-numerical month, where a dash (–) is the separator. It is right-justified within the field. **w** can be from 2 to 10, and the default is 8. It is similar to the MMDDYY. format in that if you specify the width from 2–5, the date will be truncated on the right, with SAS trying to fit as much of the year and month as possible in the space allowed. If you use 6, no dashes will be printed. At 7, SAS will print a two-digit year without a dash, and 8 or 9 will put a two-digit year before the first dash. Use a width of 10 to get a four-digit year with dashes. The following table shows the result when the date value is 14927, which corresponds to November 13, 2000:

Format Name	Result	Comment
YYMMDD4.	0011	Two-digit year, month, not enough space for day.
YYMMDD5.	00-11	Two-digit year, month, dash separator, not enough space for day.
YYMMDD6.	001113	Two-digit year, month, day, no separators.
YYMMDD7.	001113	One leading space, no separators.
YYMMDD8.	00-11-13	Two-digit year
YYMMDD9.	00-11-13	Two-digit year, leading space.
YYMMDD10.	2000-11-13	Four-digit year.

YYMMDD*xw*. Displays the date in the same way that the YYMMDD. format does, except that you can specify the separator. The **x** in the format name specifies the separator that you want to use according to the table shown here:

x	Character Displayed in Output	Comment
B	blank	
C	colon (:)	
D	dash (—)	Effectively the same as using the YYMMDD*w*. format.
N	no separator	**w** is a maximum of 8, not 10.
P	period (.)	
S	slash (/)	

The date will be right-justified within the width you specify. **w** can be from 2 to 10, and the default is 8. If you specify 2–5, the date will be truncated on the right, with SAS trying to fit as much of the day and month as possible in the space allowed. If you use 6, no separator will be used. At 7, SAS will print a two-digit year without a separator, and 8 or 9 will put a two-digit year before the first separator. The following table shows the result when the date value is 17136, which corresponds to December 1, 2006:

Format Name	Result	Comment
YYMMDDN4.	0612	Two-digit year, month, not enough space for day.
YYMMDDC5.	06:12	Two-digit year, month, colon separator, not enough space for day.
YYMMDDD6.	061201	Two-digit year, month, day, no separators.
YYMMDDP7.	061201	One leading space, no separators.
YYMMDDB8.	06 12 01	Two-digit year.
YYMMDDN8.	20061201	Because there is no separator, a four-digit year is displayed.
YYMMDDS9.	06/12/01	Two-digit year, leading space, slash separators.
YYMMDDD10.	2006-12-01	Four-digit year, dash separators, same as YYMMDD.

YYMONw.   Writes dates as a two- or four-digit year followed by the three-letter month abbreviation. It is right-justified. **w** can be from 5 to 32, and the default is 7. Use a width of 7 to get a four-digit year. If **w** is less than 7, a two-digit year will be displayed. If **w** is larger than 7, leading spaces will be added. The following table shows the result when the date value is 15323, which corresponds to December 14, 2001:

Format Name	Result	Comment
YYMON5.	01DEC	Two-digit year.
YYMON6.	01DEC	Two-digit year, 1 leading space.
YYMON7.	2001DEC	Four-digit year.
YYMON10.	2001DEC	Four-digit year, 3 leading spaces.

YYQw.   Writes date values as a two-digit or four-digit year, followed by the letter Q, and a single-digit representing the quarter of the year. It is right-justified, and **w** can be from 4 to

32. The default width is 6. Use 6 to get a four-digit year, while a width of 4 or 5 will give you a two-digit year. Specifying a width larger than 6 will only add leading spaces. The following table shows the result when the date value is 16271, which corresponds to July 19, 2004:

Format Name	Result	Comment
YYQ4.	04Q3	Two-digit year.
YYQ5.	04Q3	Two-digit year, 1 leading space.
YYQ6.	2004Q3	Four-digit year.
YYQ8.	2004Q3	Four-digit year, 2 leading spaces.

YYQxw. Writes date values as a two-digit or four-digit year, followed by a separator that you specify, and a single-digit representing the quarter of the year. **x** is the letter you use to indicate the separator according to the table shown here. Unlike the **DDMMYYxw.**, **MMDDYYxw.**, and the **YYMMDDxw.** formats, a blank is not a valid separator with this format.

x	Character Displayed in Output	Comment
C	colon (:)	
D	dash (—)	
N	no separator	*w* can be from 3–32, with a default of 4. When *w* is 3 or 4, the year will be displayed as a two-digit year.
P	period (.)	
S	slash (/)	

This format is right-justified, and **w** can be from 4 to 32. The default width is 6. Use a width 6 to get a four-digit year, while 4 or 5 will give you a two-digit year. Specifying a width larger than 6 will add leading spaces. The following table shows the result when the date value is 15253, which corresponds to October 5, 2001:

Format Name	Result	Comment
YYQN3.	014	Two-digit year, no separator.
YYQC4.	01:4	Two-digit year.
YYQS5.	01/4	Two-digit year, 1 leading space.
YYQP6.	2001.4	Four-digit year.
YYQD7.	2001-4	Four-digit year, 1 leading space.

YYQRw. Writes date values as a two-digit or four-digit year, followed by the letter Q, and the quarter of the year is represented in Roman numerals. It is right-justified, and **w** can be from 6 to 32. The default width is 8. Use 8 to get a four-digit year, while 6 or 7 will give you a two-digit year. Specifying a width larger than 8 will add leading spaces. The following table shows the result when the date value is 14099, which corresponds to August 8, 1998:

Format Name	Result	Comment
YYQR6.	98QIII	Two-digit year.
YYQR7.	98QIII	Two-digit year, 1 leading space.
YYQR8.	1998QIII	Four-digit year.
YYQR12.	1998QIII	Four-digit year, 4 leading spaces.

YYQRxw. Writes date values as a two-digit or four-digit year, followed by a separator that you specify, and the quarter of the year is displayed as a Roman numeral. **x** is the letter you use to indicate the separator according to the following table. This is another format that cannot use a blank as the separator.

x	Character Displayed in Output	Comment
C	colon (:)	
D	dash (–)	
N	no separator	*w* can be from 5–32, with a default of 7. When **w** is 5 or 6, the year will be displayed as a two-digit year.
P	period (.)	
S	slash (/)	

This format is right-justified, and **w** can be from 6 to 32. The default width is 8. Use 8 to get a four-digit year, while 6 or 7 will display a two-digit year. Specifying a width larger than 8 will add leading spaces. The following table shows the result when the date value is 17030, which corresponds to August 17, 2006:

Format Name	Result	Comment
YYQRP6.	06.III	Two-digit year.
YYQRS7.	06/III	Two-digit year, 1 leading space.
YYQRN8.	2006III	Four-digit year, 1 leading space, no separator.
YYQRC9.	2006:III	Four-digit year, 1 leading space.
YYQRD10.	2006-III	Four-digit year, 2 leading spaces.

2.4.2 Time Formats

Time formats translate seconds into one of several different ways of displaying time. Only the TIMEAMPM*w.d* and TOD*w.d* formats are specific to clock values, displaying clock values from 12:00:00 AM to 11:59:59 PM. All other formats will display hours greater than 23 when translating a value greater than or equal to 86400, which would be midnight of the following day. The display of minutes always ranges from 0 to 59, except when you are using the MMSS. format. The built-in SAS formats always display seconds from 0 to 59.

The width specification for time (and datetime) values is different from the one for date formats because it has to allow for decimal parts of seconds. Instead of **w**, time and datetime formats are specified as **w.d**, where **w** is the overall width of the entire format, and the **d** accounts for the number of digits to the right of the decimal point. **w** must be greater than (**d**+1) to account for the decimal point. As with date formats, each of these formats has its own default width specification, which is detailed in the description of the format.

HHMMw.d   Displays SAS time values as hours:minutes. It is right-justified, and does not display a leading zero in front of the hours. **w** can be from 2 to 20, with a default of 5, while **d** indicates the number of decimal places to the right of the minutes. As noted above, **w** must be greater than **d+1**, to account for the decimal point. It is different from the TIME**w.d** format in that it does not display seconds. If **d** is 0 or not present, SAS will round to the nearest minute. Otherwise, SAS will display the seconds in decimal minutes (seconds/60). The following table shows the result when the date value is 19886, which corresponds to 5 hours, 31 minutes, and 26 seconds:

Format Name	Result	Comment
HHMM2.	5	One leading space because no leading zero.
HHMM4.	5:31	No leading zero, so single-digit hours will fit.
HHMM5.	5:31	One leading space because there is no leading zero.
HHMM8.	5:31	Four leading spaces, no leading zero.
HHMM8.2	5:31.43	26 seconds = .43 minutes, one leading space, no leading zero.

HOURw.d   Displays SAS time values as hours and decimal fractions of hours. It is right-justified. **w** can be from 2 to 20, with a default of 2. **d** is the number of decimal places to the right of the hours, and **w** must be greater than **d**+1 to account for the decimal point. If you do not specify any decimal places, SAS rounds to the nearest hour. The following table shows the result when the date value is 53706, which corresponds to 14 hours, 55 minutes and 6 seconds:

Format Name	Result	Comment
HOUR2.0	15	Rounded to the nearest hour.
HOUR4.2	14.9	55 minutes, 6 seconds is .92 hours, does not leave enough space for second decimal place.
HOUR6.2	14.92	One leading space.
HOUR8.2	14.92	Three leading spaces.

MMSSw.d Displays SAS time values as minutes and seconds (mm:ss). It is right-justified. **w** can be from 2 to 20, with a default of 5. If you do not specify **w** large enough to fit minutes and seconds, SAS will round and display the minutes only. **d** will print decimal fractions (e.g., tenths or hundredths) of seconds. **w** must be greater than **d**+1 to account for the decimal point. The following table shows the result when the date value is 37269, which corresponds to the time 10:21:09 AM (10:21:09):

Format Name	Result	Comment
MMSS4.	621	Leading space.
MMSS5.	621	Two leading spaces.
MMSS8.2	621:09.0	One decimal place for tenths of seconds, because not enough space to fit. A **w** of 8 only leaves enough space for a **d** of 1 because of the decimal point.
MMSS9.2	621:09.00	Two decimal places for hundredths of seconds.

TIMEw.d Displays SAS time values as hours:minutes:seconds. It is right-justified, and does not print a leading zero in front of the hours. **w** can be from 2 to 20, with a default of 8, while **d** indicates the number of decimal places to the right of the seconds. **w** must be greater than **d**+1 to account for the decimal point. **d** will print decimal fractions (e.g., tenths or hundredths) of seconds.

This format is not restricted to a 24-hour day; if **hours** is greater than 24, then it will display the value of hours. It is different from the HHMM**w.d** format in that it does display seconds. If *d* is 0 or not present, SAS will round to the nearest second. The following table shows the result when the date value is 29794, which corresponds to the time 8:16:34 AM (8:16:34):

Format Name	Result	Comment
TIME5.	8:16	Leading space because of the single-digit hour.
TIME6.	8:16	
TIME7.	8:16:34	No leading spaces; single-digit hour allows the full time to fit in 7 spaces.
TIME8.	8:16:34	
TIME9.	8:16:34	

TIMEAMPM*w.d* Displays time in hours:minutes:seconds followed by a space and then AM or PM. It is right-justified. **w** can be from 2 to 20, and the default is 11. **w** must be greater than **d**+1, to account for the decimal point. Any time value greater than or equal to 86400 (midnight) will be displayed as the 12-hour clock time of the next day. This format does not print a leading zero. If you want the seconds to be printed, use at least 11 for the width. The following table shows the result when the date value is 11923, which corresponds to the time 3:18:43 AM:

Format Name	Result	Comment
TIMEAMPM7.	3:18 AM	Single-digit hour, 1 leading space, no seconds.
TIMEAMPM9.	3:18 AM	
TIMEAMPM11.	3:18:43 AM	Single-digit hour leaves 1 leading space.
TIMEAMPM14.	3:18:43 AM	Four leading spaces.

TODw.d Displays time in hours:minutes:seconds. It is right-justified. **w** can be from 2 to 20, and the default is 11. **d** is the decimal fraction of seconds, and must be less than **w**–1, to account for the decimal point. Any time value greater than or equal to 86400 (midnight of the next day) will be marked as the 24-hour clock time of the next day. This format does not print a leading zero. If you want the seconds to be displayed, use 8 for the width. Use at least 10 if you want decimal fractions of seconds shown. The following table shows the result when the date value is 75122, which corresponds to the time 8:52:02 PM (20:52:02):

Format Name	Result	Comment
TOD5.	20:52	
TOD8.	20:52:02	
TOD11.	20:52:02	Three leading spaces.
TOD14.	20:52:02	Six leading spaces.

2.4.3 Datetime Formats

Datetime formats translate SAS datetime values into one of several different formats. SAS datetime values are the number of seconds since midnight, January 1, 1960. You can use a format to display both the date and time. Again, you need to pay attention to the width specification in datetime formats because it allows for decimal fractions of seconds. Instead of **w**, time and datetime formats are specified as **w.d**, where **w** is the overall width of the entire format, and the **d** accounts for the number of digits to the right of the decimal point. **w** must be greater than (**d**+1) to account for the decimal point. As with date formats, each of these formats has its own default width specification, which is detailed in the description of the format.

Starting with Version 9, there are also formats that will allow you to display just the date or just the time from a datetime value, eliminating the need to use the DATEPART() or TIMEPART() functions for display purposes. Although these DT formats give the same result as their corresponding date formats, the results you get will be very different should you use a datetime format on a date value and vice versa. Datetime formats translate *seconds* since midnight, January 1, 1960, while date formats translate *days* since January 1, 1960.

The following example shows what happens when you use date formats to interpret datetime values and vice versa. If you translate the SAS date value 16838 using a date format, you will get the correct value of February 6, 2006 (❶ and ❷.) However, if you use a datetime format to translate the same value, you will get 4:40:38 AM on January 1, 1960, which corresponds to 16,838 seconds after midnight, January 1, 1960 (❸ and ❹). In similar fashion, if you translate the value 1422287527 using a datetime format, you will get 3:52:07 on January 25, 2005 (❺ and ❻.) This time, if you try to use a date format to translate this value, you will get a series of asterisks because the value is too large for the SAS date algorithm to handle (❼ and ❽.) Even if you try to get the month, day and year of this value using the appropriate functions, it will not work.

Example 2.4.1 The Difference Between Date and Datetime Values in Formats That Display Dates

```
DATA _NULL_;
date = 16838;
datetime = 1422287527;
PUT "MMDDYY10. representation of date=" date mmddyy10. /
"MONYY7. representation of date=" date monyy7. /
"DTMONYY7. representation of date=" date dtmonyy. /
"When value of date is used as a SAS *datetime* value, the date
represented is:" date datetime20. /
"DATETIME20. representation of datetime=" datetime datetime20. /
"DTMONYY7. representation of datetime=" datetime dtmonyy7. /
"MONYY7. representation of datetime=" datetime monyy7. /
"When value of datetime is used as a SAS *date* value, the date
represented is:" datetime mmddyy10.;
RUN;
```

The Log

```
DATA _NULL_;
 date = 16838;
 datetime = 1422287527;
 PUT "MMDDYY10. representation of date=" date mmddyy10. /
 "MONYY7. representation of date=" date monyy7. /
 "DTMONYY7. representation of date=" date dtmonyy. /
 "When  value  of  date  is  used  as  a  SAS *datetime* value,  the  date
 represented is:" date datetime20. //
 "DATETIME20. representation of datetime=" datetime datetime20. /
 "DTMONYY7. representation of datetime=" datetime dtmonyy7. /
"MONYY7. representation of datetime=" datetime monyy7. /
```

(continued on next page)

```
"When value of datetime is used as a SAS *date* value, the date represented
  is:" datetime mmddyy10.;
 RUN;
 MMDDYY10. representation of date=02/06/2006      ❶
 MONYY7. representation of date=FEB2006 ❷
 DTMONYY7. representation of date=JAN60 ❸
 When value of date is used as a SAS *datetime* value, the date represented
is:  01JAN1960:04:40:38      ❹

 DATETIME20. representation of datetime=  25JAN2005:15:52:07❺
 DTMONYY7. representation of datetime=JAN2005  ❻
 MONYY7. representation of datetime=*******      ❼
 When value of datetime is used as a SAS *date* value, the date represented
 is: *********      ❽
```

With that in mind, here are the formats that are applicable to datetime values.

DATEAMPM*w.d* Displays datetime values as "**ddmonyy(yy):hh:mm:ss.ss xx**", where **dd** is the day of the month, **mon** is the three-letter abbreviation for the month, and **yy(yy)** is the two- or four-digit year. A colon follows the date, and the time is represented by **hh:mm:ss.ss**, followed by a space and then AM or PM. It is right-justified. **w** can be from 7 to 40, and the default is 19. **d** is the decimal fraction of seconds, and must be less than **w**–1, to account for the decimal point. **w** must be at least 13 to print AM or PM. If **w** is 10, 11, or 12, the time is displayed as a 24-hour clock. Also, if **w–d** is less than 17, the decimal values will be truncated to fit the specified field width. This format produces two-digit years for widths of 19 or less. The following table shows the result when the date value is 1297063816.5, which corresponds to the time 7:30:17 AM on February 6, 2001:

Format Name	Result	Comment
DATEAMPM7.	06FEB01	No room for time.
DATEAMPM12.	06FEB01:07	Two leading spaces, hours is given in 24-hour clock.
DATEAMPM18.	06FEB01:07:30 AM	Two leading spaces, not enough room for seconds.
DATEAMPM18.1	06FEB01:07:30 AM	Two leading spaces, not enough room for seconds.
DATEAMPM19.	06FEB01:07:30:17 AM	
DATEAMPM19.1	06FEB01:07:30:17 AM	Not enough room for decimal portion of seconds.
DATEAMPM25.	06FEB2001:07:30:17 AM	Four leading spaces, four-digit year, and rounded seconds.
DATEAMPM25.1	06FEB2001:07:30:16.5 AM	Two leading spaces, four-digit year, fractional seconds to 1 decimal place.
DATEAMPM29.	06FEB2001:07:30:17 AM	Eight leading spaces.

DATETIME*w.d* Displays datetime values as ddmonyy(yy):hh:mm:ss.ss, where **dd** is the day of the month, **mon** is the three-letter month abbreviation, and **yy(yy)** is the two- or four-digit year. A colon follows the date, and the time is represented by **hh:mm:ss.ss**. It is similar to the DATEAMPM. format, except that it uses the twenty-four-hour clock and therefore does not display AM or PM. It is right-justified. **w** can be from 7 to 40, and the default is 19. **d** is the decimal fraction of seconds, and must be less than **w**–1 to account for the decimal point. If **w**–**d** is less than 17, the decimal values will be truncated to fit the specified field width. If **w**–**d** is less than 19, this format produces two-digit years. The following table shows the result when the date value is 1336982668, which corresponds to the time 8:04:28 AM on May 14, 2002:

Format Name	Result	Comment
DATETIME16.	`14MAY02:08:04:28`	
DATETIME18.	`14MAY02:08:04:28`	Two leading spaces.
DATETIME18.1	`14MAY02:08:04:28.0`	One decimal place.
DATETIME19.	`14MAY2002:08:04:28`	Four-digit year, not enough space for decimal point and decimal place.
DATETIME19.1	`14MAY02:08:04:28.0`	**w–d** =18, so the year is shown as a two-digit year with 1 leading space.
DATETIME20.	`14MAY2002:08:04:28`	Two leading spaces.
DATETIME20.1	`14MAY2002:08:04:28.0`	**w–d** = 19, so the year is shown as a four-digit year.
DATETIME21.	`14MAY2002:08:04:28`	Three leading spaces.
DATETIME21.2	`14MAY2002:08:04:28.00`	Four-digit year, decimal seconds to 2 places.

DTDATE*w*. Displays datetime values as the numerical day of the month, followed by the three-letter month abbreviation, and the year without any separating characters. It is right-justified within the field. **w** can be from 5–9, the default width is 7. If you want to display four-digit years, use DTDATE9. The output is identical to the output using the DATE. format. The difference is that this format will only work correctly with datetime values, while the DATE. format only works correctly with date values. The following table shows the result when the date value is 1560379389.4, which corresponds to the time 10:43:09.4 PM on June 11, 2009:

Format Name	Result	Comment
DTDATE5.	**11JUN**	
DTDATE6.	**11JUN**	Leading space, no year.
DTDATE7.	**11JUN09**	
DTDATE8.	**11JUN09**	Leading space.
DTDATE9.	**11JUN2009**	Four-digit year.

DTMONYY*w*. Displays the date from a datetime value as the three-letter month abbreviation followed immediately by the year. There are no separating characters. It is right-justified, and **w** can be from 5 to 7, with a default of 5. Specifying 5 or 6 will give you a two-digit year, while 7 will give you a four-digit year. Although this format appears to produce the same MMMyy(yy) result as the MONYY. format, the DTMONYY. format can be used only with datetime values, while the MONYY. format works only with date values. The following table shows the result 1490086128, which corresponds to the time 8:48:48 AM on March 21, 2007.

Format Name	Result	Comment
DTMONYY5.	**MAR07**	
DTMONYY6.	**MAR07**	Leading space.
DTMONYY7.	**MAR2007**	Four-digit year.

DTWKDATX*w*. Writes datetime values as day of week name, day, month-name, and year. It differs from the WEEKDATX. format in that it works on datetime values, not date values. It is right-justified, and **w** can range from 3 to 37. The default is 29, which is the maximum width of a date in this format. Specifying anything longer than 29 will cause leading spaces to be added. If the width specified is too small to display the complete day of the year and month, SAS will abbreviate. It will first abbreviate the month and then the day of the week as necessary. The following table shows the result when the date value is 1393525751.9, which corresponds to the time 6:29:11.9 PM on February 27, 2004:

Format Name	Result	Comment
DTWKDATX3.	`Fri`	
DTWKDATX11.	`Friday`	Full name of day with leading spaces.
DTWKDATX15.	`Fri, 27 Feb 04`	Leading space.
DTWKDATX16.	`Fri, 27 Feb 2004`	Four-digit year, no leading space.
DTWKDATX20.	`Friday, 27 Feb 2004`	Full name of day, month abbreviation.
DTWKDATX29.	`Friday, 27 February 2004`	Leading spaces in example, but will fit any date.
DTWKDATX33.	`Friday, 27 February 2004`	More leading spaces.

DTYEAR*w*. Displays the year for the given datetime value. DTYEAR*w*. is identical in result to the YEAR. format, but it is used with datetime values instead of date values. It is right-justified, and *w* can be from 2 to 4, with a default width of 4. When *w* is specified as 2 or 3, a two-digit year is used. The following table shows the result when the date value is 1464518782.8, which corresponds to the time 10:46:22.8 AM on May 29, 2006:

Format Name	Result	Comment
DTYEAR2.	`06`	Two-digit year.
DTYEAR3.	`06`	Two-digit year with a leading space.
DTYEAR4.	`2006`	Four-digit year.

DTYYQC*w*. Writes date values as a two-digit or four-digit year, followed by a colon, and a single-digit representing the quarter of the year. It is right-justified, and *w* can be from 4 to 6. The default width is 4. Use a width of 6 to get a four-digit year. Use 4 or 5 to get a two-digit year. This gives you the same result with datetime values as using the YYQC. format would yield with a date value. The following table shows the result when the date value is 1313917486.6, which corresponds to the time 9:04:46.6 AM on August 20, 2001:

Format Name	Result	Comment
DTYYQC4.	01:3	Two-digit year.
DTYYQC5.	01:3	Leading space, two-digit year.
DTYYQC6.	2001:3	Four-digit year.

2.5 Creating Custom Date Formats Using the VALUE Statement of PROC FORMAT

In addition to the date and time formats supplied with SAS, you can create your own custom formats with the FORMAT procedure. With dates and times, you can modify the default display of an existing SAS format, or create your own using the VALUE or the PICTURE statement. Here are two examples of modifying the default display of an existing SAS format using the VALUE statement.

Example 2.5.1 Creating Your Own Format with the VALUE Statement in PROC FORMAT

An access control company wants a report of the people whose security cards have expired as of January 1, 2005, and they have the expiration date for each card. Instead of having to read the report and determine which dates are prior to the cutoff, they want to display any date prior to January 1, 2005 as Expired. Make sure that you put the format name after the range, enclosed in brackets.

```
PROC FORMAT LIBRARY=LIBRARY;
VALUE EXP
LOW-'31DEC2004'D= "Expired"
'01JAN2005'D - HIGH=[mmddyy10.]; /* Instructs SAS to use the   */
                                 /* MMDDYY10. format for these */
                                 /* values                     */
RUN;

PROC PRINT DATA= ACCESS;
ID CARD_NUM;
VAR EXP_DATE EXP_DATE_RAW;
FORMAT EXP_DATE EXP. EXP_DATE_RAW DATE9.;
RUN;
```

CARD_NUM	EXP_DATE	EXP_DATE_RAW
84485598	11/14/2006	14NOV2006
16205371	11/27/2005	27NOV2005
63656754	01/14/2005	14JAN2005
10270040	Expired	01APR2004
94822015	Expired	04JUN2004
27800904	Expired	23OCT2004
97189418	08/14/2005	14AUG2005
70815194	03/14/2007	14MAR2007
50465401	Expired	26MAY2004
43034970	09/28/2005	28SEP2005

Example 2.5.2 Creating Your Own Format with the VALUE Statement in PROC FORMAT

In order to be able to drive the next stage in a road race, drivers must finish this stage in ten minutes or less, and the results are posted. This example shows that you can customize time formats as well as date formats. To customize datetime formats, you would specify a datetime format instead of a date or time format.

```
PROC FORMAT;
VALUE QUALIFY
LOW-'00:10:00'T=[MMSS5.]; /* Instructs the SAS System to use the */
                         /* MMSS5. format for these values       */
'00:10:00'T <- HIGH = "Did Not Qualify";
RUN;

PROC PRINT DATA=RACERS;
ID NAME;
VAR TIME;
FORMAT TIME QUALIFY.;
RUN;
```

NAME	TIME
BORK	Did Not Qualify
BOVA	Did Not Qualify
BRANTLEY	08:31
BRICKOWSKI	08:59
BURKHART	07:10
BURROUGHS	08:05
BUTLER	Did Not Qualify

2.6 Creating Custom Date Formats Using the PICTURE Statement of PROC FORMAT

To create your own date and time formats with the PICTURE statement of the FORMAT procedure, you need to use the DATATYPE= option. DATATYPE can take the DATE, TIME, or DATETIME arguments to indicate the type of value you are formatting. You then need to define your display by using one of the following date directives. These directives *are* case-sensitive. Table 2.6.1 shows the directives.

Table 2.6.1 *Picture Format Date Directives*

%a	Locale's abbreviated weekday name. Locale is defined by the LOCALE= system option.
%A	Locale's full weekday name. Locale is defined by the LOCALE= system option.
%b	Locale's abbreviated month name. Locale is defined by the LOCALE= system option.
%B	Locale's full month name. Locale is defined by the LOCALE= system option.
%d	Day of the month as a decimal number (1–31), with no leading zero. Put a zero between the percent sign and the "d" to have a leading zero in the display.
%H	Hour (24-hour clock) as a decimal number (0–23), with no leading zero. Put a zero between the percent sign and the "H" to have a leading zero in the display.
%I	Hour (12-hour clock) as a decimal number (1–12), with no leading zero. Put a zero between the percent sign and the "I" to have a leading zero in the display.

(continued on next page)

Table 2.6.1 (continued)

%j	Day of the year as a decimal number (1–366), with no leading zero. Put a zero between the percent sign and the "j" to have a leading zero in the display.
%m	Month as a decimal number (1–12), with no leading zero. Put a zero between the percent sign and the "m" to have a leading zero in the display.
%M	Minute as a decimal number (0–59), with no leading zero. Put a zero between the percent sign and the "M" to have a leading zero in the display.
%p	Locale's equivalent of either AM or PM. Locale is defined by the LOCALE= system option.
%S	Second as a decimal number (0–59), with no leading zero. Put a zero between the percent sign and the "S" to have a leading zero in the display.
%U	Week number of the year (Sunday as the first day of the week) as a decimal number (0–53), with no leading zero. Put a zero between the percent sign and the "U" to have a leading zero in the display.
%w	Weekday as a decimal number, where 1 is Sunday, and Saturday is 7.
%y	Year without century as a decimal number (0–99), with no leading zero. Put a zero between the percent sign and the "y" to have a leading zero in the display.
%Y	Year with century as a decimal number (four-digit year).
%%	The percent character (%).

NOTE: If you are going to use these directives in a picture format, you will need to add some code to handle the display of missing values.

The following example creates a format similar to the WORDDATE. format, except that leading zeros are a part of the date display. Pay special attention to how the picture is created in line 4. When you are using any of the date directives from the table above, you *must* enclose your picture string in *single* quotes. If you use double quotes, SAS will try to interpret the directives as macro calls. It will write the format to the catalog, and you will see warnings in the SAS log when the format is used. It will not work if you have a macro that has the same name as the format you created.

Example 2.6.1 Creating a Picture Format for Dates Using Date Directives

```
1  PROC FORMAT;
2  PICTURE ZWDATE
3   . - .Z = "NO DATE GIVEN"
4  LOW - HIGH = '%B %0d, %Y'  (DATATYPE=DATE);
5  RUN;
6  PROC PRINT DATA=PICTEST LABEL;
7  VAR DATE DATE2;
8  FORMAT DATE WORDDATE. DATE2 ZWDATE21.;
9  LABEL DATE = "DATE USING WORDDATE."
10 DATE2 = "DATE USING ZWDATE.";
11 RUN;
```

Line 3 defines the display for missing values. Without it, you will see the SAS missing value symbol: either a period (.) or a special missing value. Line 4 gives the picture of how the date is to be displayed. The zero preceding the day directive (%d) causes the leading zero to be printed as a part of the date.

The length of the string with the date directives determines the default width of the format. In this example, the default width is 10, but in order to make sure that all the dates print correctly, the FORMAT statement in line 8 sets the format width to 21. Following is the resulting output:

date using worddate.	date using zwdate.
June 8, 2002	June 08, 2002
October 17, 2004	October 17, 2004
May 30, 2005	May 30, 2005
.	No Date Given
November 14, 2001	November 14, 2001
March 2, 2003	March 02, 2003
.	No Date Given
April 26, 2005	April 26, 2005

Example 2.6.2 Using Date Directives and a Picture Format to Bypass a Function

This example defines a format that will display the date part of datetime values as mm/dd/yyyy. It is an alternative to using the DATEPART() function and then formatting the result. In line 3, any missing values are made to display the word Missing. The picture for all remaining values is defined in line 4. The **%0m** says that the first characters will be the numerical value of the month, with a leading zero if necessary, followed by a slash (**/**). Similarly, the **%0d** is interpreted as the numerical day of the month, again with a leading zero if necessary, followed by a slash. The picture ends with the four-digit year **%Y**. The DATATYPE= option tells the format that it will be receiving datetime values to translate. This format has a default width of ten (2 for the month, 2 for the day, 2 for the slashes, and 4 for the year.)

```
1    PROC FORMAT;
2    PICTURE avoid
3    . - .Z = "Missing"
4    LOW-HIGH = '%0m/%0d/%Y' (DATATYPE=DATETIME);
5    RUN;
6
7    DATA _NULL_;
8    sample_date = "15dec2004:3:15:00"dt;
9    PUT 'SAS datetime value = ' sample_date;
10   PUT 'SAS formatted with datetime. format =' sample_date datetime.;
11   PUT 'Using custom format avoid. =' sample_date avoid.;
12   RUN;
```

The Log

```
31    DATA _NULL_;
32    sample_date = "15dec2004:3:15:00"dt;
33    PUT 'SAS datetime value = ' sample_date;
34    PUT 'SAS formatted with datetime. format =' sample_date datetime.;
35    PUT 'Using custom format avoid. ' sample_date avoid.;
36    RUN;

SAS datetime value = 1418699700
SAS formatted with datetime. format =15DEC04:03:15:00
Using custom format avoid. =12/15/2004
```

As you can see, the only piece of the datetime value displayed with our custom format is the date. By specifying the desired pieces of the datetime value as a picture, we have eliminated the need to use the DATEPART() function. However, remember that the format AVOID. only changes the *display*. The value is still a datetime value, not a date value. If you wanted to use the actual date value (e.g., in a calculation,) you would still have to use the DATEPART() function to obtain it from the datetime value.

2.7 The PUT() Function and Formats

What happens if you need to use the formatted value of a date in a character string you're assembling? If you use the variable that contains the date value, you'll get the actual SAS date value, regardless of any permanent formats assigned to the variable. The PUT() function is used to store the formatted value of a numeric value in a character variable. The syntax is:

PUT(*value, format*);

value is a constant or a variable (either numeric or character), and **format** is the name of a SAS format. If you are formatting a character variable or constant, then the format you use must be a character format. Similarly, if you are formatting a numeric value, the format must be a numeric format. The example below demonstrates:

Example 2.7.1 Using the PUT() Function to Create a Character Date String

```
DATA _NULL_;
NUMERIC_VALUE = 17422;
B = PUT(NUMERIC_VALUE,MMDDYY10.);
PUT B=;
RUN;

B=09/13/2007   /* b is a character variable of length 10 (defined by the
                  format width) */
```

There are a couple of cautions here: first, if you've already defined the length of the character value where you store the result, it has to be at least as long as the format width.

Secondly, if you want to define the format name at run-time, you must use the **PUTN(*value,format-value*)** function, where ***value*** is either a numerical or character value (it can be a constant, variable, or a valid SAS expression), and ***format-value*** is either a character variable that contains the name of a SAS format, or a character constant that represents a format name.

CHAPTER 3

Converting Dates and Times into SAS Date, Time, and Datetime Values

In Chapter 2, I talked about performing the translation from the way SAS understands dates to the way we express them. How can we do the reverse? After all, if you have a date, time, or date and time that you need to store or manipulate, it won't be represented as a SAS date, time, or datetime value (unless it is coming from another SAS data set). The translation from common date and/or time terminology to SAS is almost as easy as going the other way, and it is done in one of two ways: the first was discussed in Section 1.4 with date, time, and datetime literals. While this works for a small number of these values that are known at compile time, how do you deal with many dates, or those that are only known at run time? By using **informats** to process them.

3.1 Avoiding the Two-Digit Year Trap

IMPORTANT

Before we get started with any details about informats, let's talk about using the YEARCUTOFF= system option as discussed in Section 1.5. Any time that you translate a date or date and time from common terminology to its SAS value, the YEARCUTOFF= system option will affect the value that is created if the term that is being translated only has two digits for the year portion. Those will be interpreted starting with the 100-year period defined in the YEARCUTOFF= system option. This rule applies to anything that is being translated into a SAS date, time, or datetime value, including date, time, or datetime literals. The example below shows how date and datetime literals are affected by the YEARCUTOFF= system option. It displays the actual SAS date or datetime value represented by the literal along with its formatted value.

Example 3.1.1 Effects of the YEARCUTOFF= System Option on Date and Datetime Literals

```
OPTIONS YEARCUTOFF=1920;
DATA _NULL_;
ARRAY a[6] a1-a6;
a1 = "23MAR2005"d;
a2 = "23MAR1905"d;
a3 = "23MAR05"d;
a4 = "19AUG1959:14:45:00"dt;
a5 = "19AUG2059:14:45:00"dt;
a6 = "19AUG59:14:45:00"dt;
DO i = 1 TO 3;
    PUT a{i}= +3 a{i}= mmddyy10.;
END;
DO i = 4 TO 6;
    PUT a{i}= +3 a{i}= datetime20.;
END;
RUN;
```

```
a1=16518          a1=03/23/2005
a2=-20007         a2=03/23/1905    /* The variable a3 has a two-digit year.
                                      The year is translated as 2005 because
                                      it falls in the span 1920-2019, and
                                      therefore, a3=a1.  */
a3=16518          a3=03/23/2005



a4=-11610900      a4=19AUG1959:14:45:00
a5=3144149100     a5=19AUG2059:14:45:00    /* The variable a6 has a two-digit
                                              year. The year is translated as
                                              1959 because it falls in the
                                              span 1920-2019, and therefore,
                                              a6=a4.  */
a6=-11610900      a6=19AUG1959:14:45:00
```

The same thing will happen when you use an informat with a date or datetime that has only two digits representing the year.

3.2 Using Informats

Informats are the opposite of formats. Where formats take values and display them in a specific fashion, informats take a series of alphanumeric characters and translate them into a single value. Dates and times are the best example of applying informats, since SAS date values are not normally how the majority of the planet expresses dates. A date such as 11/24/05, or 24-05-2002 contains non-numeric characters, so they would have to be read as character values, which is quite a distance from numeric SAS date values. As with formats, you can create (and store for future use) your own informats if one is not available within SAS to fit your needs.

To apply an informat to a variable, you use the INFORMAT statement. This will cause SAS to translate a group of characters into a value which is then stored in the variable. Informats are most commonly used with the INPUT statement as data are being read in or with the INPUT() function to translate data that are already in datasets. They are also a property in some SAS/AF objects.

You specify an informat with the informat name, followed by an optional width specification, and a period (.) Informats are like formats in that each informat has a default width that SAS will use if none is specified.

3.3 The INFORMAT Statement

The INFORMAT statement is analogous to the FORMAT statement. You use the INFORMAT statement to associate an informat with a variable in a SAS dataset. You can also remove an informat that has been permanently associated with the variable by leaving the informat name blank. You may also use the INFORMAT statement to associate an informat with a variable for the duration of the procedure (in certain procedures such as the FSEDIT procedure).

```
INFORMAT date1 mmddyy10.;  /* Any character string written into date1
will always be translated with the mmddyy10. informat throughout
SAS */

INFORMAT time3;  /* Any informat permanently associated with the
variable time3 will no longer be used to translate character
strings written into time3. */
```

3.3.1 Using Informats with the INPUT Statement

The basic syntax of an INPUT statement with an informat is:

INPUT @1 date1 mmddyy10.;

First, you will usually specify a starting column. The default starting column is 1, but you can specify the starting column with the @ sign, followed by the column number. If you do not, the starting column will be set to the current location of the input pointer. You should also specify a width for the informat to indicate how many characters are to be read. The above INPUT statement will read the first ten characters in a line, starting at the first character in a data line, and SAS will expect it to look like mmsdds(yy)yy, where **mm** is the month from 01–12, **s** represents a separator character, **dd** is a day from 01–31, and **(yy)yy** is a two- or four-digit year. The following example demonstrates that the separator character does not have to be the same on every line, and the field does not have to be exactly ten characters long.

Example 3.3.1 INPUT Statement Example

```
DATA informats_are_smart;
INPUT @1 date MMDDYY10.;
unformatted_date = date;
DATALINES;
10/17/2002
05-04-59
```

```
3-1-1940
;
RUN;

PROC PRINT;
FORMAT DATE MMDDYY10.;
RUN;
```

Obs	Date	Unformatted_date
1	10/17/2002	15630
2	05/04/1959	-242
3	03/01/1940	-7245

As you can see in the above example, the characters in each line of the DATALINES statement were converted to a SAS date value. It doesn't matter that lengths of the character strings representing date in the data are different. It is only an issue if you have more characters in the date string than the ten columns you've allocated for the date. The length of the informat must be long enough to read all of the characters in the date string.

If the characters read do not match that layout (e.g., June 26, 1994), or if the informat would yield an impossible value (e.g., February 31) SAS will set the value of the variable date1 to missing, and set the system variable _ERROR_ to 1. In general, you should know the layout of the characters before selecting an informat. In SAS Version 9, if you do not know the layout of the dates, the ANYDT family of informats (Section 3.4.4) can help.

3.3.2 Informats with the INPUT() Function

The INPUT() function is the parallel to the PUT() function, and it stores a numeric or character value as a numeric or character variable. The type of the result depends on the type of the informat that is used. A character informat (one that begins with a $) will return a character value. All of the informats used with dates, times, and datetimes are numeric; therefore, the variable returned is numeric. The syntax is:

INPUT(*character-value,informat-name*);

If you want to define the informat that is to be applied during a SAS job (at run-time), you will need to use the **INPUTN(*character-value,informat-value*)** function (or **INPUTC()**, if you want to produce a character variable) instead. ***informat-value*** represents a character variable or character constant that contains an informat name, while the INPUT() function needs an actual informat name. Make sure that you have defined the width of the informat so that it is long enough to capture all the characters in the entire character variable. The following example illustrates the use of the INPUT() function:

Example 3.3.2 INPUT() Function Example

```
DATA _NULL_;
a = "15-NOV-2003";
b = INPUT(a,DATE11.);
PUT B=;
RUN;

B=16024
```

The INPUT() function translates the date in the character variable **a** into its equivalent SAS date value and stores it in the numeric variable **b**. The DATE11. informat accounts for the length of the character variable.

3.3.3 When the Informat Does Not Match the Data Being Read

Informats, like formats, are separated into classes according to the type of data that are being read. In most cases, if you use the wrong informat for the data type, informats will return an error (set the SAS automatic variable _ERROR_ to 1), and the value of the variable being read will be set to missing.

This behavior differs from formats in that if you use the wrong type of format to display a value (e.g., if you use a date format to display a time value), no error will occur, and at worst, you will get a warning in the SAS log. However, incorrectly specifying a format will most likely cause the display to be incorrect. Example 3.3.3 shows what happens when you try to use an informat that does not match the character string that you are trying to process.

Example 3.3.3 Using the Wrong Informat

```
DATA bad_informat;
INPUT @1 date datetime18.;
DATALINES;
11-06-1988
8-25-2004
4-24-2005
;;;;
RUN;
```

The Log

```
1      DATA bad_informat;
2      INPUT @1 date datetime18.;
3      DATALINES;

NOTE: Invalid data for date in line 4 1-18.
RULE:       ----+----1----+----2----+----3----+----4----+----5----+----6----+
4          11-06-1988
date=. _ERROR_=1 _N_=1
NOTE: Invalid data for date in line 5 1-18.
5          8-25-2004
date=. _ERROR_=1 _N_=2
NOTE: Invalid data for date in line 6 1-18.
6          4-24-2005
date=. _ERROR_=1 _N_=3
NOTE: The data set WORK.BAD_INFORMAT has 3 observations and 1  variables.
NOTE: DATA statement used (Total process time):
      real time          0.53 seconds
      cpu time           0.03 seconds
7      ;;;;
8      RUN;
```

The Resulting SAS Dataset

Obs	date
1	.
2	.
3	.

As you can see in the above example, using the DATETIME. informat to process a series of character strings that do not represent datetimes produces a note in the log. It also sets the automatic variable _ERROR_ to 1 for each record where it encountered a mismatch between the informat specified and the data it attempted to read. The end result is that the value of the date variable in your output dataset is missing because SAS was not able to process the characters using the specified informat. Remember to always check your log and your dataset after reading a text file.

3.4 Listing and Discussion of Informats

Each discussion of an informat in this section will provide an explanation of the informat, its length specification and the text it is designed to process. Each section is accompanied by a table that gives examples of the text that is to be processed, along with the informat (and its length specification), and the resulting SAS date, time, or datetime value.

3.4.1 Date Informats

DATE*w*. Reads dates in the form **ddmonyy(yy)**, where **dd** represents the day of the month, **mon** is the three-letter month abbreviation, and **yy(yy)** is the two- or four-digit year. The default value of **w** is 7, but you should specify 9 if you are reading four-digit years. **dd**, **mon**, and **yy(yy)** can be separated by blanks or special characters. If you separate them, you must account for the blanks (or special characters) in the width specification. If you have blanks after the month and the day, then you need to have a width of 9 for two-digit years, or 11 for four-digit years. If the leading zero for **dd** is missing, it has no effect on the value. The following table gives examples of how to apply this informat to yield the SAS date value that corresponds to the text shown in each line.

When the characters being read are:	Use the Informat	And the result is:
20oct95	date7.	13076
8 jan 2004	date11.	16078
07-may-1960	date11.	127

DDMMYYw. Reads dates of the form **ddmmyy(yy)**, where **dd** represents the day of the month, **mm** represents the number of the month, and **yy(yy)** is the two- or four-digit year. The default value of **w** is 6, but you should specify 8 if you are reading four-digit years. **dd**, **mm**, and **yy(yy)** can be separated by blanks or special characters. If you separate them, you must account for the separating characters in the width specification. If you have blanks after the month and the day, then you need to have a width of 8 for two-digit years, or 10 for four-digit years. SAS will do its best to decipher the string if no separators are used, but some dates cannot be processed, e.g., 2112008. Without place-holding zeros or separators, there is no way to know if the date is 21 January, 2008, or 2 November, 2008. The following table gives examples of how to apply this informat to yield the SAS date value that corresponds to the text shown in each line.

When the characters being read are:	Use the Informat	And the result is:
140390	ddmmyy6.	11030
06/09/05	ddmmyy8.	16685
22-04-2003	ddmmyy10.	15817

JULIANw. Translates a Julian date in the form **YY(yy)ddd**, with the two- or four-digit year preceding the zero-filled Julian day of the year. It is right-justified. **w** can be from 5 to 32, and the default is 5. If you specify 5, the year portion of the Julian date is two digits long. If you specify 7 or more, the year portion is four digits long. Zeros must fill the space between the year and day values; for example, the fifth day of the year must be given as 005. Any date preceding the year 1582 on the Gregorian calendar cannot be read as a Julian value. The following table gives examples of how to apply this informat to yield the SAS date value that corresponds to the text shown in each line.

When the characters being read are:	Use the Informat	And the result is:
77284 (October 11, 1977)	julian5.	6493
2004005 (January 5, 2004)	julian7.	16075
2002111 (April 21,2002)	julian10.	15451

MMDDYY*w.* Reads dates of the form **mmddyy(yy)**, where **mm** represents the number of the month, **dd** represents the day of the month, and **yy(yy)** is the two- or four-digit year. The default value of **w** is 6, but you should specify 8 if you are reading four-digit years. **dd**, **mm**, and **yy(yy)** can be separated by blanks or special characters. If you separate them, you must account for the blanks in the width specification. If you have blanks after the month and the day, then you need to have a width of 8 for two-digit years, or 10 for four-digit years. SAS will do its best to decipher the string if no separators are used, but some dates cannot be processed, e.g., 1272003. Without place holding zeros or separators, there is no way to know if the date is January 27, 2003, or December 7, 2003. The following table gives examples of how to apply this informat to yield the SAS date value that corresponds to the text shown in each line.

When the characters being read are:	Use the Informat	And the result is:
041705	mmddyy6.	16543
1/15/2004	mmddyy10.	16085
08281996	mmddyy10.	13389

MONYY*w.* Reads dates of the form **monyy(yy)**, where **mon** is the three-letter month abbreviation, and **yy(yy)** is the two- or four-digit year. Using this informat will set the SAS date value that corresponds to the first day of the month. The default value of **w** is 5, but you should specify 7 if you are reading four-digit years. The following table gives examples of how to apply this informat to yield the SAS date value that corresponds to the text shown in each line.

When the characters being read are:	Use the Informat	And the result is:
JAN05	monyy5.	16437 (January 1, 2005)
dec1920	monyy7.	-14275 (December 1, 1920)
aug2020	monyy7.	22128 (August 1, 2020)

PDJULG4. Reads a packed Julian date in hexadecimal format for IBM computers. The width specification is always 4, because the Julian date is parsed as follows: the four-digit Gregorian year is written in the first two bytes, and the three-digit integer that represents the day of the year is in the next one-and-a-half bytes. The last half-byte contains all binary 1's, which indicates the value is positive. There is no example given for this informat because packed decimal Julian dates yield non-printable characters.

PDJULIw. Also reads a packed Julian date in hexadecimal format for IBM computers. It differs from the PDJULG. informat in that it expects the two digits of the century in the first byte, followed by two digits of the year in the second byte. The next one-and-a-half-bytes store the three-digit integer that corresponds to the day of the year, while the last half-byte is filled with hexadecimal 1's, representing a positive number. The century and year are calculated by subtracting 1900 from the four-digit Gregorian year. Once again, there is no example, since packed decimal Julian dates yield non-printable characters.

YYMMDDw. Reads dates of the form **yy(yy)mmdd**, where **yy(yy)** is the two- or four-digit year, **mm** represents the number of the month, and **dd** represents the day of the month. The default value of **w** is 6, but you should specify 8 if you are reading four-digit years. **yy(yy)**, **mm**, and **dd** can be separated by blanks or special characters. If you separate them, you must account for the separating characters in the width specification. If you have blanks after the month and the day, then you need to have a width of 8 for two-digit years, or 10 for four-digit years. The following table gives examples of how to apply this informat to yield the SAS date value that corresponds to the text shown in each line.

When the characters being read are:	Use the Informat	And the result is:
041205	yymmdd6.	16410
20030227	yymmdd8.	15763
20030227	yymmdd10.	15763
1978-07-11	yymmdd10.	6766

YYMMN*w*. Reads dates of the form yy(yy)mm, where ***yy(yy)*** is the two- or four-digit year, and ***mm*** represents the number of the month. The day is automatically set to 1. The default value of ***w*** is 4, but you should specify 6 if you are reading four-digit years. The N in the informat name is necessary. You may not use any separating characters between the month and the year. This informat will produce a date value that is equal to the first day of the month given. The following table gives examples of how to apply this informat to yield the SAS date value that corresponds to the text shown in each line.

When the characters being read are:	Use the Informat	And the result is:
7805	yymmn4.	6695 (May 1, 1978)
200504	yymmn6.	16527 (April 1, 2005)
199211	yymmn6.	11993 (November 1, 1992)

YYQ*w*. Reads dates of the form yy(yy)Qq, where ***yy(yy)*** is the two- or four-digit year followed by the letter Q and ***q*** is a number from 1 to 4, indicating the quarter of the year. The date value produced by this informat will correspond to the first day of the given quarter. Use 6 for ***w*** if you are reading four-digit years, or 4 if you are reading two-digit years. The default ***w*** in Version 6 is 4, while Versions 7 and above have a default ***w*** of 6. The following table gives examples of how to apply this informat to yield the SAS date value that corresponds to the text shown in each line.

When the characters being read are:	Use the Informat	And the result is:
04Q1	yyq4.	16071 (January 1, 2004)
2000Q3	yyq6.	14792 (July 1, 2000)
1985Q2	yyq6.	9222 (April 1, 1985)
2003Q4	yyq6.	15979 (October 1, 2003)

3.4.2 Time Informats

MSEC8. Reads IBM mainframe time values accurate to the nearest millisecond. The width is 8 because the OS TIME macro and STCK system instructions store their time values in 8 bytes.

PDTIME4. Converts packed decimal time values contained in SMF and RMF records produced by IBM mainframe systems to SAS time values. The width is shown as 4 because SMF and RMF records are 4 bytes long.

RMFDUR4. Converts IBM SMF duration records into SAS time values. The width is shown as 4 because SMF records are 4 bytes long.

STIMER*w*. Reads times produced by the STIMER system option in the SAS log. This informat has no default width. It reads times and interprets them based on colons and decimal points. If there is one colon, the first two digits are minutes, and the last two are seconds. If there are two colons, the digits preceding the first colon are hours, the next set of two digits are minutes, and the last two are seconds. If there is a decimal point, the value following the decimal point is translated as a decimal fraction of seconds. It can read time values in the following formats, where **hh** corresponds to hours, **mm** corresponds to minutes, **ss** corresponds to seconds, and **ff** corresponds to decimal fractions of seconds:

ss
ss.ff
mm:ss
mm:ss.ff
hh:mm:ss
hh:mm:ss.ff

```
DATA _NULL_;
INPUT A STIMER11.;
PUT A;
DATALINES;
33          ❶
51.60       ❷
14:05       ❸
3:11.03     ❹
1:19:21     ❺
11:46:17.74   ❻
RUN;
```

Corresponding SAS Time Values	
❶	33
❷	51.6
❸	845
❹	191.03
❺	4761
❻	42377.74

TIMEw. This will read times in the form hh:mm:ss.ff, where **hh** indicates the hours, **mm** is minutes, and **ss** is the number of seconds. **ff** indicates decimal fractions of seconds. Both seconds and their decimal fractions are assumed to be zero if they are not present. This informat can read AM and PM time values. If **hh** is greater than 24, and/or **mm** and **ss** are greater than 60, the time value read will give the correct number of seconds, even if it is greater than 86399.99 (the number of seconds in a day.) It will parse the time value according to the number of colons in the input string as shown in the table below.

hh
hh:mm
hh:mm:ss
hh:mm:ss.ff

The program below demonstrates how the informat works, with the SAS time values that correspond to the input line on the right hand side.

```
DATA _NULL_;
INPUT A TIME10.;
PUT A;
DATALINES;
124:46       ❶
14:11:03.3   ❷
1:27 PM      ❸
6:30 AM      ❹
18:53        ❺
RUN;
```

Corresponding SAS Time Values	
❶	449160
❷	51063.3
❸	48420
❹	23400
❺	67980

TU4. Converts IBM mainframe timer units to SAS time values. It is used when reading IBM mainframe timer values under other operating systems. The width is 4 because the OS TIME macro returns a 4-byte word.

3.4.3 Datetime Informats

DATETIME*w*. This reads SAS datetime values. The datetime value must be in the form **ddmonyy(yy)**, followed by a blank or a special character, and then the time in the format **hh:mm:ss.ff**. *dd* represents the day of the month, *mon* is the three-letter month abbreviation, and *yy(yy)* is the two- or four-digit year. *hh* indicates the number of hours, *mm* is the number of minutes, and *ss* is the number of seconds. *ff* indicates fractional parts of seconds. Both seconds and fractional seconds are assumed to be zero if they are not present. *w* can be from 13 to 40, with a default of 18.

NOTE

If you use a two-digit year, SAS will apply the YEARCUTOFF= system option in translating the year. This informat can also read AM and PM time values.

```
DATA _NULL_;
INPUT A DATETIME22.;
PUT A;
DATALINES;
22APR2004 5:23 PM       ❶
     22APR2004-17:23       ❷
     22APR2004:05:23:15 PM   ❸
     22APR2004/17:23:15.6    ❹
     RUN;
```

Corresponding SAS Datetime Values	
❶	1398273780
❷	1398273780
❸	1398273795
❹	1398273795.6

3.4.4 ANYDT and Its Variants

The release of SAS 9 has addressed a problem with the processing of dates and times that has plagued SAS since the beginning. While informats handle the translation of a string of characters into SAS date and time values, in order to use them you had to know what the string of characters looked like before you processed them. Add to that the many ways that dates and times are represented, and you wind up with the potential for error, and using more than a few PUT _INFILE_ statements over the years. There is now a series of three informats that will intelligently and, for the most part, successfully enable you to avoid this problem.

CAUTION

The potential for confusion exists with DDMMYY, MMDDYY, and YYMMDD values, especially in the presence of two-digit-year values. Remember that the SAS system option DATESTYLE (detailed in Section 1.5) indicates how such confusions will be resolved. Once again, the possible values for the DATESTYLE= system option are shown in Table 3.4.1.

Table 3.4.1 *Values for the DATESTYLE= system option*

MDY	Sets the default order as month, day, year. "11-01-06" would be translated as November 1, 2006.	**YDM**	Sets the default order as year, day, month. "11-01-06" would be translated as June 1, 2011.
MYD	Sets the default order as month, year, day. "11-01-06" would be translated as November 6, 2001.	**DMY**	Sets the default order as day, month, year. "11-01-06" would be translated as January 11, 2006.
YMD	Sets the default order as year, month, day. "11-01-06" would be translated as January 6, 2011.	**DYM**	Sets the default order as day, year, month. "11-01-06" would be translated as June 11, 2001.
LOCALE (default)	Sets the default value according to the LOCALE= system option. When the default value for the LOCALE= system option is "English_US", this sets DATESTYLE to MDY. Therefore, by default, "11-01-06" would be translated as November 1, 2006.		

The ANYDTDTE., ANYDTDTM., and ANYDTTME. informats will translate dates, datetime values, and time values, respectively, into their corresponding SAS values. This translation will be performed without having to know the representation of these dates, datetime, and time values in advance. There are limits to the types of representations these informats will process, and using these informats will take more CPU time than if you used one of the regular informats to process your data.

IMPORTANT

ANYDTDTE*w.* This will translate data that can be read with the following informats: DATE, DATETIME, DDMMYY, JULIAN, MMDDYY, MONYY, TIME, YYMMDD, or YYQ into SAS date values. Note that it can extract date values from a datetime value. However, if only a time value is given, *the date is assumed to be January 1, 1960.* **w** can range from 5 to 32, and the default width is 9.

The following program illustrates the use and function of this informat. The table that follows the program was created from the output file from the program. The table details each of the character strings that are used as input from the DATALINES statement in the program.

```
OPTIONS DATESTYLE=MDY;
DATA _NULL_;
FILE "ANYDTDTE.TXT";
RETAIN TAB '09'X;
INPUT A $20. @
   @1 B ANYDTDTE20.;   /* RE-READ SAME DATA LINE INTO A NUMERIC VARIABLE
                          USING THE ANYDTDTE. INFORMAT */
PUT  A TAB  B TAB  B WORDDATE.;
DATALINES;
05172004
20040517
2004Q1
051704
17052004
170504
17MAY2004:15:12:06
15:12:06
2004138
MAY2004
17MAY2004
;
run;
```

Input String	SAS Date Value	Formatted Value	Notes
05172004	16208	May 17, 2004	
20040517	16208	May 17, 2004	
2004Q1	16071	January 1, 2004	
051704	16208	May 17, 2004	
17052004	.	.	The input string cannot be translated reliably regardless of the order of precedence. The date could be April 20, 1705 (04/20/1705), or May 17, 2004. SAS will try to apply the MMDDYY. informat because of the DATESTYLE option in effect, but 17 is an invalid value for month.
170504	20943	May 4, 2017	As opposed to the example above with a four-digit year, this can be translated. However, this is greatly affected by the DATESTYLE option in effect. See the special caution in Example 3.4.1 at the end of this section on the ANYDT informats.
17MAY2004:15:12:06	16208	May 17, 2004	
15:12:06	0	January 1, 1960	Time values are translated as seconds after midnight, 1/1/1960, so the date is 1/1/1960.
2004138	16208	May 17, 2004	Julian date
MAY2004	16192	May 1, 2004	

ANYDTDTMw. This will translate data that can be read with the following informats: DATE, DATETIME, DDMMYY, JULIAN, MMDDYY, MONYY, TIME, YYMMDD, or YYQ, and create SAS datetime values. If only a time value is given, *the date is assumed to be January 1, 1960.* **w** can range from 1 to 32, and the default width is 19. The following table uses the same input data as is used in the ANYDTDTE. informat example above.

Input String	SAS Datetime Value	Formatted Value
05172004	1400371200	17MAY04:00:00:00
20040517	1400371200	17MAY04:00:00:00
2004Q1	1388534400	01JAN04:00:00:00
051704	1400371200	17MAY04:00:00:00
17052004	.	The input string cannot be translated reliably regardless of the order of precedence.
170504	1809475200	As opposed to the example above with a four-digit year, this can be translated. However, this is greatly affected by the DATESTYLE option in effect. See the special caution in Example 3.4.1 at the end of this section on the ANYDT informats.
17MAY2004:15:12:06	1400425926	17MAY04:15:12:06
15:12:06	54726	01JAN60:15:12:06
2004138	1400371200	17MAY04:00:00:00
MAY2004	1398988800	01MAY04:00:00:00
17MAY2004	1400371200	17MAY04:00:00:00

ANYDTTMEw. This will take data that can be read with the DATE, DATETIME, DDMMYY, JULIAN, MMDDYY, MONYY, TIME, YYMMDD, or YYQ informats and translate it to SAS time values. ANYDTTMEw. can obtain time values from a datetime value. However, if only a date value is given, *the time is assumed to be 12:00 AM*. **w** can range from 1 to 32, and the default width is 8. This table uses the same input data that was used in the ANYDTDTE. informat example.

Input String	SAS Time Value	Formatted Value
05172004	0	12:00:00AM
20040517	0	12:00:00AM
2004Q1	0	12:00:00AM
051704	0	12:00:00AM
17052004	.	The input string cannot be translated reliably regardless of the order of precedence.
170504	0	12:00:00AM
17MAY2004:15:12:06	54726	3:12:06PM
15:12:06	54726	3:12:06PM
2004138	0	12:00:00AM (Julian date)
MAY2004	0	12:00:00AM
17MAY2004	0	12:00:00AM

Example 3.4.1 Be Careful When You Use the ANYDT. Series of Informats

The ANYDT. informats are not perfect. Let's look a little more closely at the interaction between the DATESTYLE= system option and the ANYDTDTE. informat. As noted in the above examples, the characters 17052004 cannot be reliably translated, so SAS returns a missing value regardless of what DATESTYLE= option is in effect. However, it was also shown that 170504 can be translated. The SAS value you get for those characters varies widely depending on the DATESTYLE= option.

When OPTIONS DATESTYLE=myd, '170504' is translated as: . (missing)

When OPTIONS DATESTYLE=ymd, '170504' is translated as: 20943 (05/04/2017)

When OPTIONS DATESTYLE=ydm, '170504' is translated as: . (missing)

When OPTIONS DATESTYLE=dmy, '170504' is translated as: 16208 (05/17/2004)

When OPTIONS DATESTYLE=dym, '170504' is translated as: . (missing)

When OPTIONS DATESTYLE=mdy, '170504' is translated as: 20943 (05/04/2017)

Three of the DATESTYLE settings yield missing values, while two yield the expected date values. But for some reason, the MDY setting (italics) gives you the value you'd expect from the YMD setting, and this may not be what you want.

While it is always a good idea to check all data that you are converting to SAS from another source and especially when you are converting dates, times, and date-time values, it is *critical* if you are using the ANYDT. informats.

Date and Time Functions

SAS has several functions to manipulate dates, times, and datetime values. The functions can be categorized according to what they do. You can obtain the current date, time, or datetime (as specified by the computer's clock). You can also easily extract pieces of dates, times or datetimes as numerical values from their corresponding SAS values, or you can assemble SAS date, time, and datetime values from SAS variables or constants. Another set of functions operates with intervals such as weeks or months.

4.1 Current Date and Time Functions

Current date and time functions have no arguments, and return SAS values as noted in the following table. The values are obtained from the operating system's clock.

DATE(), TODAY()	These functions are identical, and they both return the current date as a SAS date value.
TIME()	This returns the current time as a SAS time value.
DATETIME()	This returns the current date and time as a SAS datetime value.

4.2 Extracting Pieces from SAS Date, Time, and Datetime Values

The extraction functions all use a single argument (represented by *arg* in the following table), which represents a SAS date, time, or datetime value. This can be either a SAS variable name or the appropriate constant. If two-digit year values are used, the result will be subject to the YEARCUTOFF= option value in effect. They all return a numeric value as the result. The following table gives examples of how to apply each function, along with relevant comments for each example:

Function Name	Explanation	Example
DATEPART(*arg*)	Extracts the date from a SAS datetime value as a SAS date value.	DATEPART('21MAR1998:17:07:00'dt) = 13959 (March 21, 1998)
DAY(*arg*)	Extracts the number of the day of the month from a SAS date value.	DAY("14OCT2008"d) = 14
HOUR(*arg*)	Extracts the hour from a SAS time value.	HOUR("7:35:00"t) =7
JULDATE(*arg*)	Extracts the Julian date from a SAS date value. It will return a four- or five-digit value if the year portion of the date falls within the 100-year span defined by the YEARCUTOFF= option. In order to be Y2K compliant, you should use the JULDATE7(); function.	JULDATE("09MAY2004"d) = 4130 (Result is returned as a numeric value, so there are no leading zeros.) JULDATE("09MAY1890"d) = 1890129
JULDATE7(*arg*)	Extracts the Julian date with a four-digit year from a SAS date value. This is the Y2K-compliant version since it always returns a seven-digit number, regardless of the year.	JULDATE7("09MAY2004"d) = 2004130 JULDATE7("09MAY1890"d) = 1890129
MINUTE(*arg*)	Extracts the minutes from a SAS time value.	MINUTE("12:17:43 PM"t) = 17
MONTH(*arg*)	Extracts the numerical month from a SAS date value.	MONTH("22AUG2005"d) = 8
QTR(*arg*)	Extracts the quarter of the year from a SAS date value.	QTR("8JAN2000"d) = 1
SECOND(*arg*)	Extracts the seconds from a SAS time value.	SECOND("2:17:43"t) = 43

(*continued on next page*)

Function Name	Explanation	Example
TIMEPART(*arg*)	Extracts the time portion from a SAS datetime value as a SAS time value.	TIMEPART("06SEP1976:13:36:33"dt) = 48993 (1:36:33 PM)
WEEK(*arg*)	Extracts the week number from a SAS date value, where Sunday is the first day of the week.	WEEK("02JAN2005"d) = 1
(Version 9.1.3) WEEK(*arg,descriptor*)	Extracts the week number from a SAS date value. *descriptor* can be "U", "V", or "W" (case-insensitive), and it refers to the algorithm used to calculate the first week of the year. The U algorithm calculates weeks based on Sunday being the first day of the week. The V algorithm calculates weeks to the ISO standard. Monday is the first day of the week, and the first week of the year is defined as the one that contains both January 4 and the first Thursday of the year. The W algorithm calculates weeks based on Monday being the first day of the week without restriction.	WEEK("02JAN2005"d,"U") = 1 January 2, 2005 was a Sunday, so the first week of the year has started. WEEK("02JAN2005",d,"V") = 53 This week is defined as being the 53rd week in 2004, because it doesn't contain the first Monday or Thursday of the year. WEEK("02JAN2005"d,"W") = 0 The year 2005 has started, but weeks are calculated with Monday as the first day of the week. Therefore, the first week of 2005 doesn't start until January 3, 2005, so this is week 0 of 2005.
WEEKDAY(*arg*)	Extracts the number of the day of the week, where Sunday=1, Monday=2, etc. from a SAS date value.	WEEKDAY("14APR1999"d) = 4 (Wednesday, April 14, 1999)
YEAR(*arg*)	Extracts the year from a SAS date value. If you use a date constant (as in the example) and not a SAS date value, it is important to remember that the YEARCUTOFF option affects two-digit years.	If OPTIONS YEARCUTOFF=1920; YEAR("19JUL10"d) = 2010 YEAR("19JUL1910"d) = 1910

4.3 Creating Dates, Times, and Datetimes from Numbers

This series of functions will create SAS date, time, and datetime values from numerical variables or constants.

DATEJUL(julian-date); creates a SAS date value from a numeric value representing a Julian date. **julian-date** must be of the type **yy(yy)ddd,** where **yy(yy)** is two or four digits representing the year, and **ddd** must be between 1–365 (366 if a leap year.) If you use two digits for the year, the YEARCUTOFF= option will be used to determine the century. The following table gives examples of how to apply this function, along with relevant comments for each example:

Sample Function Call	SAS Date Value	Formatted with MMDDYY10. Format	Comments
OPTIONS YEARCUTOFF=1920; DATEJUL(21286)	–13959	10/13/1921	With the YEARCUTOFF value of 1920, the 21 is interpreted as 1921.
OPTIONS YEARCUTOFF=2000; DATEJUL(21286)	22566	10/13/2021	If YEARCUTOFF is 2000, the 21 is interpreted as 2021.
DATEJUL(2005174)	16610	06/23/2005	
DATEJUL(1989005)	10597	01/05/1989	
DATEJUL(00368)	.	.	368 is not a valid value for a Julian day, so the function returns a missing value.

DHMS(date,hour,minute,second); creates a SAS datetime value. All four arguments are required. **date** is a SAS date value, which can be either a numeric value or a date constant. If you use a two-digit year in a date constant, the date will be translated according to the YEARCUTOFF= option. **hour** is a numeric variable or constant, **minute** is a numeric variable or constant, and **second** is a numeric variable or constant. **Hour, minute** and **second** are not restricted to their clock times; therefore, **hour** can be greater than 24, while **minute** and **second** can be greater than 60. The following table gives examples of how to apply this function, along with relevant comments for each example:

Sample Function Call	Datetime Value	Formatted with DATETIME. Format	Comments
DHMS("08JUN1995"d,15,24,0)	1118244240	08JUN1995:15:24:00	
DHMS("10SEP1999"d,3,0,0)	1252551600	10SEP1999:03:00:00	
DHMS("02FEB2004"d,11,54,15)	1391342055	02FEB2004:11:54:15	
DHMS("30JUN1992"d,8,7,93)	1025510913	30JUN1992:08:08:33	The value 93 is just an argument. The function ultimately returns the datetime value in seconds and the formatted value converts the result. Therefore, 93 seconds becomes 1 minute, 33 seconds, which adds 1 to the minute value of 7, and reduces the seconds to 33.
DHMS("26APR2003"d,23,0,28)	1367017228	26APR2003:23:00:28	

HMS(hour,minute,second); creates a SAS time value. **hour** is a numeric variable or constant, **minute** is a numeric variable or constant, and **second** is a numeric variable or constant. All three arguments must be present. **hour, minute,** and **second** are not restricted to their clock times, so that **hour** can be greater than 24, while **minute** and **second** can be greater than 60. The following table gives examples of how to apply this function, along with relevant comments for each example:

Sample Function Call	Time Value	Formatted with TIME. Format	Formatted with TIMEAMPM. Format	Comments
HMS(18,0,9)	64809	18:00:09	6:00:09 PM	
HMS(7,45,80)	27980	7:46:20	7:46:20 AM	The time is not displayed as "7:45:80" because the value is returned as the total number of seconds, and the format is applied to that. Neither the TIME. nor the TIMEAMPM. formats display minute or second values greater than 59.
HMS(15,03,35.56)	54215.56	15:03:36	3:03:36 PM	
HMS(8,17,33)	29853	8:17:33	8:17:33 AM	
HMS(21,14,28)	76468	21:14:28	9:14:28 PM	

MDY(month,day,year); creates a SAS date value from the arguments. All three arguments are required. **month** is a numeric variable or constant, **day** is a numeric variable or constant, and **year** is a numeric variable or constant. If year is two digits, the century will be determined by the YEARCUTOFF= option. If a value given for any of the arguments is not valid, such as MDY(2,31,2004) (February 31, 2004), the function will return a missing value and give you an "invalid argument to function" message in the log. The following table gives examples of how to apply this function, along with relevant comments for each example:

Sample Function Call	SAS Date Value	Formatted with MMDDYY10. Format
MDY(9,3,1876)	–30434	09/03/1876
MDY(12,14,15)	20436	12/14/2015
MDY(3,26,1915)	–16352	03/26/1915
MDY(5,22,2033)	26805	05/22/2033
MDY(1,7,2004)	16077	01/07/2004

YYQ(year,qtr);   creates a SAS date value from the arguments. Both arguments are required. year is a numeric variable or constant representing the year, and **qtr** is a numeric variable or constant between 1 and 4, representing the quarter of the year. If year is two digits, the century will be determined by the YEARCUTOFF= system option. This function returns the date of the first day of the quarter in the given year. The following table gives examples of how to apply this function, along with relevant comments for each example:

Sample Function Call	SAS date Value	Formatted with MMDDYY10. Format	Comment
YYQ(1995,1)	12784	01/01/1995	
YYQ(99,3)	14426	07/01/1999	YEARCUTOFF=1920
YYQ(25,2)	–12693	04/01/1925	YEARCUTOFF=1920, so year is 1925.
YYQ(2005,2)	16527	07/01/2005	

4.4 Calculating Intervals

Calculating an interval can sometimes be done by using simple math. However, SAS provides functions to calculate intervals because, in most cases, simple math is only an approximation. The function is going to be more accurate. For example, one of the mathematical equations for calculating age is (current date−date of birth)/365.25. This approximation uses the .25 to account for leap years. In addition, some functions provide more capability, such as the ability to redefine the starting and ending dates of periods of time such as years. SAS interval functions are very powerful and can solve a host of problems.

4.4.1 Calculating Elapsed Time with DATDIF() and YRDIF()

DATDIF()

The DATDIF() function calculates the number of days between two dates. The syntax is:

> **DATDIF(start,end,basis);**

Start is the starting date, which can be a date constant, a numeric variable or a SAS expression. ***End*** is the ending date, also a date constant, a numeric variable, or a SAS expression. ***Basis*** is a character constant or variable that tells SAS how to calculate the difference. It has two possible values:

'30/360', which sets each month to 30 days, and the year to 360 days, regardless of how many days are in each month or year in the span between the two dates. If a day is at the end of a month (e.g., February 28/29 or March 31) it will be considered as the 30th of the month.

'ACT/ACT', which uses the actual number of days in each month and year in the span between the two dates. This is the default, and it is identical to subtracting ***start*** from ***end***.

If you use a character constant for ***basis***, remember that it will need to be enclosed in quotes, or you will get an error.

Example 4.4.1 The DATDIF Function

```
DATA _NULL_;
a = DATDIF('19JUL2002'd,'19JUL2003'd,'30/360');
b = DATDIF('19JUL2002'd,'19JUL2003'd,'act/act');
PUT a= / b=;
RUN;
```

The Log

```
a=360
b=365
```

The value of A is 360 because basis is "30/360", indicating a year of 360 days by definition. The value of b is 365, because it was calculated with the actual number of days in the years 2002–2003. If 2003 had been a leap year, then the value would be 366.

YRDIF()

The YRDIF function calculates the number of years between two dates. It is generally more accurate than using mathematical approximation[1]. The syntax is:

YRDIF(*start, end, basis*);

start is the starting date, which can be a date constant, a numeric variable or a SAS expression. *end* is the ending date, also a date constant, a numeric variable, or a SAS expression. *basis* is a character constant or variable that tells SAS how to calculate the difference. It has four possible values, as compared with the two possibilities in the DATDIF function:

'30/360', which sets each month to 30 days, and the year to 360 days, regardless of how many days are in each month or year in the span between the two dates. If a day is at the end of a month (e.g., February 28/29 or March 31) it will be considered as the 30th of the month.

'ACT/ACT', which uses the actual number of days in each month and year in the span between the two dates. This is the default.

---

[1] For details about age calculations, see http://support.sas.com/sassamples/quicktips/calcage.html.

'ACT/360', which uses the actual number of days between the two dates to calculate the number of years, but it uses a 360-day year, regardless of how many days are in each year, so the result is number of days divided by 360.

'ACT/365', which uses the actual number of days between the two dates to calculate the number of years, but uses a 365-day year, regardless of how many days are in each year, so the result is number of days divided by 365.

The following example shows the effect that each *basis* definition has on the value that the YRDIF() function returns given the same period of time.

Example 4.4.2: The YRDIF() Function

```
DATA _NULL_;
A = YRDIF('07AUG1963'D,'08MAY2005'D,'30/360');
B = YRDIF('07AUG1963'D,'08MAY2003'D,'ACT/ACT');
C = YRDIF('07AUG1963'D,'08MAY2003'D,'ACT/360');
D = YRDIF('07AUG1963'D,'08MAY2003'D,'ACT/365');
PUT '30 DAY MONTH, 360 DAY YEAR = ' A;
PUT 'ACTUAL DAYS IN MONTH, ACTUAL YEAR = ' B;
PUT 'ACTUAL DAYS IN MONTH, 360 DAY YEAR = ' C;
PUT 'ACTUAL DAYS IN MONTH, 365 DAY YEAR = ' D;
RUN;
```

The Log

```
30 day month, 360 day year = 41.752777778
actual days in month, actual year = 39.750684932
actual days in month, 360 day year = 40.330555556
actual days in month, 365 day year = 39.778082192
```

As you can see, all four results are different, and this is due to the way they were calculated. If each month is 30 days long, and the year is 360 days long, there are 41.75 years between the two dates. If the actual days in a month are used, but the years are standardized to 360 days, there are 40.33 years between the dates.

If the actual days and actual year are used for the calculation, then the value is 39.75. If you use the actual days, but define the year to be 365 days long, the value is 39.77, a discrepancy caused by the leap years during that time span. The final example below illustrates the difference between using the YRDIF() function, the INTNX() function (since it counts elapsed intervals) and mathematical estimation.

4.4.2 The Basics of SAS Intervals

SAS has several interval definitions that are used with dates, times, and datetimes. Although the standard interval definitions handle many standard time intervals, you have the option of using interval multipliers and interval shift arguments, which allow you to define intervals. Multipliers and shift intervals are discussed in detail in Section 5.3. Table 4.4.1 is a list of all the standard interval definitions and the periods that they describe:

Table 4.4.1 *SAS Interval Definitions Used with Dates, Times, and Datetimes*

Category	Interval Name	Definition	Default Starting Point
Date	DAY	Daily intervals	Each day
	WEEK	Weekly intervals of seven days	Each Sunday
	WEEKDAYdaysW	Daily intervals with Friday-Saturday-Sunday counted as the same day (five-day work week with a Saturday-Sunday weekend). *days* identifies the individual numbers of the weekend day(s) by number (1=Sunday ... 7=Saturday). By default, days="17", so the default interval is WEEKDAY17W.	Each day

(continued on next page)

Table 4.4.1 *(continued)*

Category	Interval Name	Definition	Default Starting Point
Date	TENDAY	Ten-day intervals (a U.S. automobile industry convention)	1st, 11th, and 21st of each month
	SEMIMONTH	Half-month intervals	First and sixteenth of each month
	MONTH	Monthly intervals	First of each month
	QTR	Quarterly (three-month) intervals	1-Jan 1-Apr 1-Jul 1-Oct
	SEMIYEAR	Semi-annual (six-month) intervals	1-Jan 1 Jul
	YEAR	Yearly intervals	1-Jan
Datetime	DTDAY	Daily intervals	Each day
	DTWEEK	Weekly intervals of seven days	Each Sunday
	DTWEEKDAYdaysW	Daily intervals with Friday-Saturday-Sunday counted as the same day (five-day work week with a Saturday-Sunday weekend). days identifies the individual weekend days by number (1=Sunday ... 7=Saturday). By default, days="17", so the default interval is DTWEEKDAY17W.	Each day
	DTTENDAY	Ten-day intervals (a U.S. automobile industry convention)	1st, 11th, and 21st of each month
	DTSEMIMONTH	Half-month intervals	First and sixteenth of each month

(continued on next page)

Table 4.4.1 *(continued)*

Category	Interval Name	Definition	Default Starting Point
Datetime	DTMONTH	Monthly intervals	First of each month
	DTQTR	Quarterly (three-month) intervals	1-Jan 1-Apr 1-Jul 1-Oct
	DTSEMIYEAR	Semiannual (six-month) intervals	1- Jan 1 Jul
	DTYEAR	Yearly intervals	1-Jan
	DTSECOND	Second intervals	Seconds
	DTMINUTE	Minute intervals	Minutes
	DTHOUR	Hour intervals	Hours
TIME	SECOND	Second intervals	Seconds
	MINUTE	Minute intervals	Minutes
	HOUR	Hourly intervals	Hours

4.4.3 Interval Calculation Functions: INTCK and INTNX

The interval calculation functions INTCK() and INTNX() utilize SAS interval definitions. INTCK() finds the number of intervals between two given dates, times, or datetimes. In contrast, INTNX() finds the date, time, or datetime that results after a given number of intervals have been applied to an initial date, time, or datetime value.[2]

---

[2] As of SAS 9, there are other interval calculation functions available with SAS/ETS and/or SAS High Performance Forecasting. Releases beyond SAS 9.1 might add some holiday-related functions as well.

INTCK()

The INTCK() function counts the number of intervals between two dates, times, or datetimes. It does so by counting from the beginning of the given interval at *start-of-period* and the beginning of the interval at *end-of-period*. ***interval*** is the SAS designation for a period of time, and can be a character literal or character variable that corresponds to one of the defined time intervals (see Table 3.4.1.) The syntax of the INTCK function is:

$$\text{INTCK}(\textit{interval}, \textit{start-of-period}, \textit{end-of-period});$$

In essence, the INTCK() function is counting the number of times that the period *interval* begins between *start-of-period* and *end-of-period*, inclusive. It does *not* count the number of complete intervals between *start-of-period* and *end-of-period*. This also means that the count does not begin with *start-of-period*, but at the beginning of the first interval after that. The following sample logs demonstrate how INTCK() counts. As an example, take the dates Saturday, December 31, 2005 and Sunday, January 1, 2006.

The Log

```
137   DATA _NULL_;
138   v1 = INTCK('DAY','31dec2005'd,'01jan2006'd);
139   v2 = INTCK('WEEK','31dec2005'd,'01jan2006'd);
140   v3 = INTCK('MONTH','31dec2005'd,'01jan2006'd);
141   v4 = INTCK('YEAR','31dec2005'd,'01jan2006'd);
142   PUT v1= +3 v2= +3 v3= +3 v4= +3;
143   RUN;


v1=1     v2=1     v3=1     v4=1
```

All of the intervals are equal to 1 even though only one day has passed! January 1, 2006 is the start of day, week, month, and year intervals. The starting day occurred on December 31, and the ending day began on January 1. That's obvious. However, Sunday is the beginning of the week; therefore, the week for December 31 started on Sunday, December 25. Sunday, January 1 is the beginning of the next week, so one week has elapsed between the start of the two weeks containing both dates, and that causes v2 to be equal to 1. The month for December 31 started on December 1, and the month for January 1, 2006 started on January 1, so one MONTH interval has elapsed between the start of the intervals for the two dates, and therefore, v3 is 1. Finally, the year for December 31, 2005 started on January 1 of 2005, while the year for January 1, 2006 starts on the same date, which causes v4 to be 1.

All of the values are equal to 1 because the INTCK() function is counting the DAY, WEEK, MONTH, and YEAR interval boundary which occurs because of the number of days, weeks, months or years that have passed.

To complete the picture of how INTCK() works, let's look at the effect that the ending date has on INTCK().

The Log

```
144   DATA _NULL_;
 145   v5 = INTCK('DAY','31dec2005'd,'06jan2006'd);
 146   v6 = INTCK('WEEK','31dec2005'd,'06jan2006'd);
 147   v7 = INTCK('MONTH','31dec2005'd,'06jan2006'd);
 148   v8 = INTCK('YEAR','31dec2005'd,'06jan2006'd);
 149   PUT v5= +3 v6= +3 v7= +3 v8= +3;
 150   RUN;

 v5=6     v6=1     v7=1     v8=1
```

Here you can see that although 6 days have elapsed, still only 1 week, month, and year have elapsed according to INTCK()! That is because the start of the week, month, and year interval for January 6, 2006, is still January 1, 2006, and that is what INTCK() is counting. Here are some more examples of the use of INTCK():

Example 4.4.3 The INTCK Function – the Basics

INTCK(**'DAY'**, '15jun2003'd,'22jun2003'd) = **7**
INTCK(**'WEEK'**, '01jan2001'd, '01jan2002'd) = **52**
INTCK(**'DTDAY'**, '01oct1872:08:00:00'dt ,'20dec1872:18:00:00'dt) = **80**
INTCK(**'MONTH'**, '05mar1978'd, '01may1978'd) = **2**

Example 4.4.4 The INTCK Function – Counting Backwards

INTCK(**'YEAR'**, '22dec2005'd, '16jul2000'd) =**–5**

In this example, the *from* date is after the *to* date, and therefore, the answer is negative. Since INTCK() counts interval boundaries, the answer is –5 because it starts counting at the start of a year. The start of the year for December 22, 2005 is January 1, 2005, so that is where the count begins. January 1, 2004; January 1, 2003; January 1, 2002; January 1, 2001; and January 1, 2000 are the beginning dates of the years that it is counting.

Example 4.4.5 The INTCK Function – Counting Weekdays

A) INTCK(**'WEEKDAY17W'**,'08JAN2006'd,'10JAN2006'd) = **2**

B) INTCK(**'WEEKDAY17W'**,'09JAN2006'd,'10JAN2006'd) = **1**

C) INTCK(**'WEEKDAY17W'**,'06JAN2006'd,'07JAN2006'd) = **0**

If you are counting the number of work weekdays that have elapsed, you must be careful to remember that INTCK() counts interval boundaries, and that the starting date is not counted in the answer. A) above seems perfectly reasonable. You would expect that there would be two weekdays between Sunday, January 8, 2006 and Tuesday, January 10, 2006. However, you might think that there are two weekdays in the span Monday, January 9, 2006 to Tuesday, January 10, 2006, but the INTCK() function counts only 1 weekday, because it starts counting with Tuesday. Why is C) equal to zero then? January 7, 2006 is a Saturday, and since we've defined the weekend days as Saturday and Sunday, the WEEKDAY17W interval does not begin until Monday, so no interval boundaries have been passed.

The last example for INTCK() illustrates the difference between the three methods that have been discussed for calculating elapsed years using SAS: the YRDIF() function, the INTCK() function (because it counts elapsed intervals), and mathematical estimation.

Example 4.4.6: The YRDIF() Function as Opposed to Mathematical Estimation and INTCK()

This uses the same dates, August 7, 1963 and May 8, 2005 for all three methods, but the first line uses the YRDIF() function with "**ACT/ACT**", while the mathematical approximation divides the number of days between the two dates by 365.25. While the discrepancy is minute in this example, the difference is caused by the fact that the number of leap years in the period (11) is not divisible by 4, rendering the value 365.25, an approximation. The INTCK function counts interval boundaries from their beginning, so it is counting the number of January firsts between January 1, 1963, and January 1, 2003. In effect, unless you were born on January 1, using INTCK() to calculate your age will make you old before your time!

```
DATA _NULL_;
b = YRDIF('07aug1963'd,'08may2003'd,'ACT/ACT');
e = ('08may2003'd - '07aug1963'd)/365.25;
g = INTCK('YEAR','07aug1963'd,'08may2003'd);
PUT 'actual days in month, actual year = ' b;
PUT 'math approximation = ' e;
PUT 'Using INTCK = ' g;
RUN;
```

The Log

```
actual days in month, actual year = 39.750684932
math approximation = 39.750855578
Using INTCK = 40
```

INTNX()

The INTNX() function takes a given SAS date, time, or datetime value, and calculates a new value based on a given number of intervals. Where INTCK() calculates the number of intervals between any two date, time, or datetime values, INTNX() takes the start of the period and increments it by a number of intervals to give the end-of the period. The syntax is:

INTNX(*interval,start-from,number-of-increments,alignment*);

interval is one of the SAS intervals defined in Table 3.1, and can be a character literal or character variable that evaluates to one of the defined intervals. ***start-from*** is the starting date, time, or datetime value, which can be a constant, numeric variable, or a SAS expression. ***number-of-increments*** is an integer constant or a numeric variable that indicates how many intervals to advance. If it is not an integer, only the integer portion of the value will be used.

alignment sets the returned date, time, or datetime value according to one of four predefined settings. The function calculates the dates at the beginning of the interval period, and then the alignment argument adjusts the result. The values are Beginning or B, Middle or M, and End or E. The default value for ***alignment*** is Beginning. The Sameday or S alignment operator was added in SAS 9. The Sameday argument cannot be used with the DTQTR, DTSEMIYEAR, or the DTYEAR intervals, and it has no effect on time values.[3]

The next series of examples demonstrates various uses and effects of the INTNX() function. The first example is the default use of the function with each of the date, time, and datetime intervals, while the second shows what happens when you use a non-integer as the increment to the function. Example 4.4.9 illustrates the use of the alignment arguments, and Example 4.4.10 shows how to use the alignment arguments to yield specific dates.

Example 4.4.7 The INTNX() Function with Default Alignment

Each of the examples below increments the date, time, or datetime value by 3 of the interval shown in bold italics. For dates and datetimes, the start date is the same, Thursday, November 2, 2000. The only difference is that datetime intervals (interval names starting with DT) return datetime values, not dates. The interval values for datetime values are calculated in seconds, not days. By default, the INTNX() function returns the beginning of the interval given. If you wish to change this, use the alignment argument discussed above.

---

[3] Check releases beyond SAS 9.1 to find whether the SAME operator replaces the SAMEDAY operator.

INTNX(**'DAY'**,'11/02/2000'd,3) =**11/05/2000**	
INTNX(**'DTDAY'**,'02NOV2000:15:00:00'dt,3) = **05NOV2000:00:00:00**	The time returned is midnight of 11/05/2000, not 3 p.m.
INTNX(**'WEEKDAY17W'**,'11/02/2000'd,3) = **11/07/2000**	The returned date is the following Tuesday. Saturday (7) and Sunday (1) don't count since they are defined as the weekend days.
INTNX(**'DTWEEKDAY17W'**,'02NOV2000:15:00:00'dt,3) = **07NOV2000:00:00:00**	
INTNX(**'WEEK'**,'11/02/2000'd,3) =**11/19/2000**	The value returned is the Sunday of the week, not 21 calendar days.
INTNX(**'DTWEEK'**,'02NOV2000:15:00:00'dt,3) = **19NOV2000:00:00:00**	
INTNX(**'TENDAY'**,'11/02/2000'd,3) = **12/01/2000**	The value returned is the first of the month, 29 calendar days, not 30. This interval will always return either the 1st, 11th, or 21st of the month.
INTNX(**'DTTENDAY'**,'02NOV2000:15:00:00'dt,3) = **01DEC2000:00:00:00**	
INTNX(**'SEMIMONTH'**,'11/02/2000'd,3) = **12/16/2000**	Although November has 30 days, the returned date is the 16th, not 45 calendar days, which would be the 17th.
INTNX(**'DTSEMIMONTH'**,'02NOV2000:15:00:00'dt,3) = **16DEC2000:00:00:00**	

INTNX(**'MONTH'**,'11/02/2000'd,3) = **02/01/2001**	The date returned is the beginning of the following month, not the same date. To get the same date, use the "S" (same day) alignment argument.
INTNX(**'DTMONTH'**,'02NOV2000:15:00:00'dt,3) = **01FEB2001:00:00:00**	
INTNX(**'QTR'**,'11/02/2000'd,3) = **07/01/2001**	The first day of the quarter is returned.
INTNX(**'DTQTR'**,'02NOV2000:15:00:00'dt,3) = **01JUL2001:00:00:00**	
INTNX(**'SEMIYEAR'**,'11/02/2000'd,3) = **01/01/2002**	The beginning of the 3rd semi-year after 11/02/2000 is January 1, 2002.
INTNX(**'DTSEMIYEAR'**,'02NOV2000:15:00:00'dt,3) = **01JAN2002:00:00:00**	
INTNX(**'YEAR'**,'11/02/2000'd,3) = **01/01/2003**	The beginning of the 3rd year after 11/02/2000 is January 1, 2003.
INTNX(**'DTYEAR'**,'02NOV2000:15:00:00'dt,3) = **01JAN2003:00:00:00**	
INTNX(**'SECOND'**,'8:00:00 AM't,3) = **8:00:03 AM**	
INTNX(**'MINUTE'**,'8:00:00 AM't,3) = **8:03:00 AM**	
INTNX(**'HOUR'**,'8:00:00 AM't,3) = **11:00:00 AM**	

Example 4.4.8 The INTNX() Function – Using Non-Integer Increments

INTNX(**'HOUR '**, '16:45't ,2.5) **= 18:00:00**

When *number-of-increments* is not an integer, SAS will take the integer part as the value. In this case, 2.5 becomes 2. The beginning of the first hour increment is 17:00, and the second is 18:00. Since *alignment* is the default value of B or *beginning*, that sets the answer to the beginning of the hour, which gives the result of 18:00.

Example 4.4.9 The INTNX() Function with Alignment Arguments

INTNX(**'WEEK'**,'11/02/2000'd,3,'B') **=11/19/2000**	Beginning of the week.
INTNX(**'WEEK '**,'11/02/2000'd,3,'M ')**=11/22/2000**	Middle of the interval.
INTNX(**'WEEK '**,'11/02/2000'd,3,'S')**=11/23/2000**	Same day, so the interval ends on the same day of the week (Wednesday), leaving the duration at 21 days.

Example 4.4.10 The INTNX() Function with Alignment Arguments to Give Specific Dates

Set the date to the beginning of the week.	INTNX(**'WEEK'**,'27JUL2002'd,0,'B') = July 21, 2002
Advance to the start of the 3rd semi-month period.	INTNX(**'SEMIMONTH'**,'01MAY1998'd,3,'B') = June 16, 1998
Advance to the end of the month.	INTNX(**'MONTH'**,'05MAR2005'd,1,'E') = April 30, 2005
Advance to the end of the quarter after next.	INTNX(**'QTR'**,'06NOV1996'd,2,'E') = June 30, 1997

CHAPTER 5

Deeper into Dates and Times with SAS

5.1 Macro Variables and Dates

There is a high potential for confusion when it comes to the subject of macro variables and dates. Although you have access to dates and times in the SAS macro language with the automatic macro variables &SYSDATE, &SYSDATE9, &SYSDAY, and &SYSTIME, the display of these values is fixed, and therefore they do not give you the power of SAS formats, or of SAS date and time functions.

5.1.1 Automatic Macro Variables

None of these automatic macro variables may be assigned values with %LET or CALL SYMPUT(), or by any other means. They are read-only, and work by reading the operating system clock when a SAS session is started. This means that they do not change within a given SAS session.

&SYSDATE This automatic macro variable cannot be modified, and displays the system date as a SAS date value formatted with the DATE7. format. If the date according to the operating system when the SAS session starts is July 17, 2004, then &SYSDATE would be equal to "17JUL04". This example shows the value of &SYSDATE when the date is October 4, 2002:

```
/* The computer's date is October 4, 2002 */
3   %PUT &SYSDATE;

04OCT02
```

&SYSDATE9 This automatic macro variable is identical to &SYSDATE, except that it formats the system date with the DATE9. format, and is therefore Y2K-compliant. If the operating system date is November 22, 2005, then &SYSDATE9 would be equal to "22NOV2005". This example shows the value of &SYSDATE9 when the date is October 4, 2002:

```
/* The computer's date is October 4, 2002 */
6   %PUT &SYSDATE9;

04OCT2002
```

&SYSDAY This automatic variable displays in English the name of the day that the SAS session began (according to the operating system). If you started a SAS job or session on Wednesday, February 11, 1998, &SYSDAY would be equal to "Wednesday". This example shows the value of &SYSDAY when the date is Friday, October 4, 2002:

```
/* THE COMPUTER'S DATE IS OCTOBER 4, 2002 */
9   %PUT &SYSDAY;

Friday
```

&SYSTIME   This automatic variable displays the system time (according to the operating system) in TIME5. format. If the system time is 3:36 PM when you start SAS, then &SYSTIME would be equal to "15:36". This example shows the value of &SYSTIME when the time is 10:36 PM:

```
/* THE CURRENT TIME IS 10:36 PM */
12  %PUT &SYSTIME;

22:36
```

5.1.2 Putting Dates into Titles

One of the prime uses of dates in macro variables is for custom titles and footnotes. To use a macro variable in a title, you just need to enclose the TITLE or FOOTNOTE statement that contains the macro variable with double quotes like this:

TITLE "This is how to put a macro variable in a title: Today's Date is &SYSDATE9";

To see this in context, look at the following program:

```
TITLE "This is how to put a macro variable in a title: Today's Date is
&SYSDATE9";
ODS RTF FILE="examples\title.rtf";
PROC PRINT DATA=sashelp.company (OBS=5);
ID job1;
VAR depthead;
RUN;
ODS RTF CLOSE;
```

Here is the resulting output:

This is how to put a macro variable in a title: Today's Date is 21JUN2005

JOB1	DEPTHEAD
MANAGER	1
ASSISTANT	2
ACCOUNTANT	2
MANAGER	1
ADMIN	2

While the automatic macro variables &SYSDATE, &SYSDATE9, &SYSDAY, and &SYSTIME may give you the information you need, their formats are fixed and cannot be changed. Given the multiple ways that dates, times, and datetimes can be displayed, it would be good if you could use the formats and functions to get the display of dates and times that you want in your titles.

5.1.3 Using %SYSFUNC() to Create Dates, Times, and Datetimes in Macro Variables

Many of the date and time functions, as well as formats, are available in the SAS macro language through the %QSYSFUNC() and %SYSFUNC() macro functions. When you are using one of these macro functions with dates and times, it has two arguments: the first is the date, time, or datetime function you wish to use. The second argument is optional, but it is the format that should be applied to the result.

Example 5.1.1 Date Functions with %QSYSFUNC()

This example puts the formatted value of today's date into the macro variable &date, and demonstrates the difference between unformatted and formatted dates in macro variables. The date value in &RAWDATE is stored in macro space as a text string, and will need to be converted if you want to use it for any calculations. &DATE shows that the WORDDATE. format right-justifies the date display. So, in order to remove the leading spaces, the %LEFT() macro function is used and the result is stored in the macro variable &DATEJ.

```
7    %LET rawdate=%QSYSFUNC(DATE());
8    %LET date=%QSYSFUNC(DATE(),WORDDATE32.);
9    %LET datej=%LEFT(%QSYSFUNC(DATE(),WORDDATE32.));
10   %PUT &rawdate;
16592                        /* SAS Date Value */
11   %PUT &date;
                             June 5, 2005  /* Formatted */
12   %PUT &datej;
June 5, 2005  /* Formatted and left-justified */
```

Now you can use the result in a title or footnote.

Example 5.1.2 Taking a Macro Text String and Turning It into a SAS Date Value

---

This example takes the macro value from the automatic macro variable &SYDATE9 and will get the SAS date value from it. The PUTN() function is used here instead of the PUT() function because the format is determined during execution of the statement.

```
13 %LET DATEVAL = %SYSFUNC(PUTN("&SYSDATE9"D,5.)); /* USE PUTN(), NOT
                                                       PUT()! */

14 %PUT sysdate is &sysdate9;
sysdate is 05JUN2005
15 %PUT The SAS date value is: &dateval;
The SAS date value is: 16592
```

5.1.4 Using CALL SYMPUT() and SYMGET() with Dates, Times, and Datetimes

The CALL SYMPUT() and SYMGET() functions are also used to communicate between the DATA step and the macro world. CALL SYMPUT() takes a value and stores it in macro space from within a DATA step, while SYMGET() takes a macro variable and allows you to use it in a DATA step. The difference between run-time and compile time is very important. You cannot use a %LET statement to store macro values that are defined during execution of a DATA step. You also cannot use a macro variable reference (with an ampersand (&)) in the same DATA step where the macro variable is created with CALL SYMPUT(). The next example uses what was discussed in Sections 5.1.2 and 5.1.3 to show how you can obtain a date value from a SAS data set and then put it in a macro variable to use in a title.

Example 5.1.3 Using CALL SYMPUT to Create a Macro Variable Containing a Formatted SAS Date

This example calculates the most recent date in a given variable from a dataset, and then creates a macro variable containing the formatted value to be used in a title.

```
/* Get the maximum date in the variable */
PROC MEANS DATA=BOOK.PMDATA MAX NOPRINT;
VAR LAST_DATE;
OUTPUT OUT=TMP MAX=MAX;
RUN;

/*Transfer it to a macro variable using CALL SYMPUT() */
DATA _NULL_;
SET TMP;
CALL SYMPUT('ASOF',LEFT(PUT(MAX,WORDDATE32.)));
RUN;

OPTIONS NODATE;  /* REMOVE AUTOMATIC SYSTEM DATE ON PAGE */

/*CREATE TITLE with macro variable &asof */
TITLE "PROJECT DATA LAST UPDATED ON &ASOF";

ODS RTF FILE="EXAMPLES\SYMPUT.RTF";
PROC PRINT DATA=BOOK.PMDATA LABEL;
VAR LAST_DATE;
RUN;
```

This is the resulting report:

Project Data Last Updated on June 1, 2005

Obs	Date last modified
1	10/29/2003
2	08/13/2003
3	10/31/2003
4	.
5	*06/01/2005*
6	03/30/2005
7	*06/01/2005*
8	*06/01/2005*
9	07/21/1997
10	07/23/1997
11	12/03/2001

Since the maximum date in the data is June 1, 2005 (highlighted in bold italics for the example,) that is the value that is formatted and passed to the CALL SYMPUT function as the macro variable &asof.

5.2 Shifting SAS Date and Time Intervals

The interval function **INTCK**(*interval,start-of-period,end-of-period*) counts the number of intervals between two given SAS dates, times, or datetimes, while the function **INTNX**(*interval,start-from,number-of-increments,alignment*) advances a given date, time, or datetime by a specified number of intervals. Both functions use the standard SAS interval definitions (see Table 5.2.1). However, the INTCK() function counts at the start point of a given interval, while the INTNX() function advances to the beginning of the given interval. The ***alignment*** argument available with INTNX() will adjust the *result* to the middle, end, or—if you're using Version 9—same day of the interval. Nonetheless, the intervals are still measured from their starting point, and the adjustment is only applied *after* INTNX() has moved the specified number of intervals.

This leads to problems when your definition of an interval doesn't exactly match the standard SAS definition of that same interval. For example, the standard SAS definition of year has the first day of the year defined as starting January 1. What happens if your fiscal year starts on July 1, and that is how you wish to measure your year? What if your semi-month period is truly 14 days long?

Shift operators and interval multipliers allow you to do this. You have the ability to define the start of an interval definition by adding a shift indicator to it. The shift indicator defines the number of shift points to move the start of the interval. The table below shows the SAS interval name and what the shift increment period is.

Table 5.2.1 *SAS Intervals and their Shift Points*

CATEGORY	INTERVAL	DEFINITION	SHIFT POINT
DATE	**DAY**	Daily intervals	Days
	WEEK	Weekly intervals of seven days	Days
	WEEKDAY<daysW>	Daily intervals with Friday-Saturday-Sunday counted as the same day. **days** identifies the individual weekend days by number (1=Sunday ... 7=Saturday). By default, days="17", so the default interval is **WEEKDAY17W.**	Days
	TENDAY	Ten-day intervals	Ten day periods
	SEMIMONTH	Half-month intervals	Semimonthly periods
	MONTH	Monthly intervals	Months
	QTR	Quarterly (three-month) intervals	Months
	SEMIYEAR	Semi-annual (six-month) intervals	Months
	YEAR	Yearly intervals	Months

(continued on next page)

Table 5.2.1 (*continued*)

CATEGORY	INTERVAL	DEFINITION	SHIFT POINT
DATETIME	DTDAY	Daily intervals	Days
	DTWEEK	Weekly intervals of seven days	Days
	DTWEEKDAY<daysW>	Daily intervals with Friday-Saturday-Sunday counted as the same day. **days** identifies the individual weekend days by number (1=Sunday ... 7=Saturday). By default, days="17", so the default interval is **DTWEEKDAY17W.**	Days
	DTTENDAY	Ten-day intervals	Ten day periods
	DTSEMIMONTH	Half-month intervals	Semimonthly periods
	DTMONTH	Monthly intervals	Months
	DTQTR	Quarterly (three-month) intervals	Months
	DTSEMIYEAR	Semiannual (six-month) intervals	Months
	DTYEAR	Yearly intervals	Months
	DTSECOND	Second intervals	Seconds
	DTMINUTE	Minute intervals	Minutes
	DTHOUR	Hour intervals	Hours
TIME	SECOND	Second intervals	Seconds
	MINUTE	Minute intervals	Minutes
	HOUR	Hourly intervals	Hours

For our first example, let's use the standard case of a fiscal year that starts on July 1. To measure a year interval in those terms, you would have to move the start of the year by seven months. The start point is always included in the count of months to shift. The shift value is added to the interval name by appending it with a decimal point. Therefore, in order to advance the start of the YEAR interval seven months to July 1, you would use the interval name "YEAR.7", as illustrated by the following:

Example 5.2.1 Moving the Start of a Year Interval from January 1 to July 1

```
DATA _NULL_;
a = INTCK('YEAR','01jan2003'd,'06jul2004'd);
b = INTCK('YEAR.7','01jan2003'd,'06jul2004'd);
PUT 'INTCK with YEAR interval=' a + 3 'INTCK with YEAR.7 interval=' b;
RUN;
```

The Log

```
6     DATA _NULL_;
7     a = INTCK('YEAR','01jan2003'd,'06jul2004'd);
8     b = INTCK('YEAR.7','01jan2003'd,'06jul2004'd);
9     PUT 'INTCK with YEAR interval=' a +3 'INTCK with YEAR.7 interval=' b;
10    RUN;

INTCK with YEAR interval=1    INTCK with YEAR.7 interval=2
```

Since INTCK() counts the start of interval boundaries, when the interval is "YEAR", it is measuring from January 1, 2003 until January 1, 2004 (the start of the year containing July 6, 2004). When the interval is "YEAR.7", you have shifted the beginning of the YEAR interval by 7 months, which declares that the year starts on July 1. In the example above, that shift causes the measurement to commence on July 1, 2002 (the start of the year containing January 1, 2003) and end on July 1, 2004, which is the start of the year containing July 6, 2004. That is why the value returned is 2, not 1.

While shift operators allow you to move the starting point of any given SAS interval, an interval multiplier allows you to define the length of your own intervals. You define a custom interval by applying a multiplier value to an interval. For example, if you want to measure biweekly (14-day) periods, you would take the WEEK interval, and multiply it by 2. This makes your interval name "WEEK2", and it measures 14-day periods starting on Sunday. Example 5.2.2 demonstrates the creation and use of a custom interval, by using a multiplier of 2 for the WEEK interval. It also illustrates some of the features of custom intervals.

Example 5.2.2 Using an Interval Multiplier to Create a Custom Interval

```
DATA _NULL_;
a = INTNX('WEEK','08feb2004'd,2);
b = INTNX('WEEK2','08feb2004'd,2);
PUT 'INTNX with WEEK interval=' a weekdate32.;
PUT 'INTNX with WEEK2 interval=' b weekdate32.;
RUN;
```

The Log

```
68    DATA _NULL_;
69    a = INTNX('WEEK','08feb2004'd,2);
70    b = INTNX('WEEK2','08feb2004'd,2);
71    PUT 'INTNX with WEEK interval=' a weekdate32.;
72    PUT 'INTNX with WEEK2 interval=' b weekdate32.;
73    RUN;

INTNX with WEEK interval=        Sunday, February 22, 2004
INTNX with WEEK2 interval=          Sunday, March 7, 2004
```

In the above example, the start of the week interval two weeks from February 8 is February 22, so it makes sense that the start of the second biweekly period from February 8 is March 7. This is not as simple as it seems. When using SAS intervals, you must always keep in mind that intervals are always measured from the beginning of the starting interval to the beginning of the ending interval. This is independent of the starting and ending dates that you supply. Let's move the starting date in Example 5.5.2 backward by one week to February 1, 2004.

The Log

```
1    DATA _NULL_;
2    a = INTNX('WEEK','01feb2004'd,2);
3    b = INTNX('WEEK2','01feb2004'd,2);
4    PUT 'INTNX with WEEK interval=' a weekdate32.;
5    PUT 'INTNX with WEEK2 interval=' b weekdate32.;
6    RUN;

INTNX with WEEK interval=        Sunday, February 15, 2004
INTNX with WEEK2 interval=          Sunday, February 22, 2004
```

Advancing by two WEEKS (line 2) is still a difference of 14 days, as expected. But what happened in line 3? Shouldn't you get 28 days instead of 21? No, because the start of the WEEK2 interval containing February 1, 2004 is January 25, 2004, and that is where the INTNX() function begins its count. Here is a sample program to produce the starting dates of the intervals:

```
DATA tricky;
DO intervals= 0 TO 5;
   b = INTNX('WEEK2','01jan2004'd,intervals);
   OUTPUT;
END;
RUN;
PROC PRINT DATA=tricky LABEL;
ID interval;
VAR b;
FORMAT b weekdate32.;
LABEL interval="Number of WEEK2 intervals from January 1, 2004"
      b="Starting Date of Interval"
;
RUN;
```

The resulting table (below) shows the starting dates for the first five WEEK2 intervals of 2004. If the date(s) supplied fall between the starting dates of any two boundaries, the interval count (or incrementing) will commence from the starting date of the previous interval.

Number of WEEK2 intervals from January 1, 2004	Starting Date of Interval
0	Sunday, December 28, 2003
1	Sunday, January 11, 2004
2	Sunday, January 25, 2004
3	Sunday, February 8, 2004
4	Sunday, February 22, 2004
5	Sunday, March 7, 2004

How does SAS determine what the starting date of a given interval is if you use a multiplier? It takes the multiplied interval you've created and starts counting beginning with January 1, 1960. This is true for all multiplied intervals except multiplied WEEK intervals. Multiplied WEEK intervals are counted starting from Sunday, December 27, 1959, because weeks are defined as starting on Sundays, and January 1, 1960 was a Friday.

You may also use a multiplier and a shift indicator together if needed. Interval multipliers are directly appended to the interval name (e.g., WEEK2), and shift indicators are appended to the interval name with a leading decimal point (e.g., WEEK.5). To use both the multiplier and shift operator, you first append the multiplier to the interval name to create a new interval name (e.g., WEEK2), and then you append the shift indicator to the interval name. Given a multiplier of 2, and a shift of 5, the interval name becomes "WEEK2.5".

To demonstrate, let's expand on the example used in 5.2.2. What if you wanted your biweekly periods to start on January 1, 2004? January 1, 2004 was a Thursday, so you want to move the starting date to the fifth day of the week (Sunday=1, therefore, Thursday=5.) Now we'll generate the same table as in Example 5.2.2, using a shift indicator of 5 days in addition to the biweekly interval of WEEK2.

Example 5.2.3 Using Both an Interval Multiplier and Shift Operator to Create a Custom Interval

```
DATA tricky;
DO interval= 0 TO 5;
    b = INTNX('WEEK2.5','01jan2004'd,interval);
    OUTPUT;
END;
RUN;
PROC PRINT DATA=tricky LABEL;
ID interval;
VAR b;
FORMAT b weekdate32.;
LABEL interval="Number of WEEK2.5 intervals from January 1, 2004"
      b="Starting Date of Interval"
;
RUN;
```

Here's the resulting output:

Number of WEEK2.5 intervals from January 1, 2004	Starting Date of Interval
0	Thursday, January 1, 2004
1	Thursday, January 15, 2004
2	Thursday, January 29, 2004
3	Thursday, February 12, 2004
4	Thursday, February 26, 2004
5	Thursday, March 11, 2004

When you use a multiplier, you also have the ability to define your shifts within the entire interval created by the multiplier. As an example, let's create a decade interval by using YEAR10 as the interval. Remember that intervals start at the beginning of the boundary, so the decades would start at the beginning of the first year of the decade. What can you do if you want to define the decade as starting in May of the middle year in the decade (e.g., May of 1955 as opposed to January of 1950?)

To shift intervals across years, you need to use the first nested interval within the YEAR interval, which is MONTH. So you would use (*number of years to shift* *12) to calculate the number of months you need to shift. If you wanted to shift 5 years, you would use 5*12=**60**. Add 5 to that, which shifts the starting month from January to May, and your interval definition is now YEAR10.65. That would be decades starting in May of years that end in 5. The code below shows the effect of moving the interval by 65 months.

```
DATA tricky;
DO interval= 0 TO 5;
   b = INTNX('YEAR10','01sep1950'd,interval);
   c = INTNX('YEAR10.65','01sep1950'd,interval);
   OUTPUT;
END;
RUN;
```

```
PROC PRINT DATA=tricky LABEL;
ID interval;
VAR b c;
FORMAT b c weekdate32.;
LABEL interval="Number of intervals from January 1, 1950"
      b="Starting Date of YEAR10 Interval"
      c="Starting Date of YEAR10.65 Interval"
;
RUN;
```

For this table, we show the unshifted result alongside the shifted result for comparison:

Number of intervals from January 1, 1950	Starting Date of YEAR10 Interval	Starting Date of YEAR10.65 Interval
0	Sunday, January 1, 1950	***Tuesday, May 1, 1945***
1	Friday, January 1, 1960	Sunday, May 1, 1955
2	Thursday, January 1, 1970	Saturday, May 1, 1965
3	Tuesday, January 1, 1980	Thursday, May 1, 1975
4	Monday, January 1, 1990	Wednesday, May 1, 1985
5	Saturday, January 1, 2000	Monday, May 1, 1995

The first thing to note is that even though the date we specified is September 1, the starting date of the interval is January 1, because that is the start of the YEAR interval. In the second column, you can see that the interval has been shifted. Even though the starting date of the YEAR10. interval is January 1, 1950, the shifted interval itself starts on Tuesday, May 1, 1945 (italics), not May 1, 1955. Why? Because it is the start of the interval that contains January 1, 1950.

The most important thing to remember about using intervals, multipliers, and shift operators is that all intervals, no matter how they are defined, are measured from their beginning regardless of what date(s) you provide when you are using them. The alignment arguments 'B', 'M', 'E', and 'S' for the INTNX() function do not adjust the date until *after* the function has executed and calculated its start-of-interval result (also remember that these alignment arguments only work with the INTNX() function).

No matter what, you may use the interval multipliers and shift operators to move the starting point of an interval, and you may use them anywhere that you can use a date, time, or datetime interval.

5.3 Graphing Dates

When you use dates, times, or datetime values in SAS/GRAPH, you have to remember that they are numbers. This has a large impact on the axes and labeling. Graphing a result over a period of time without using formats or intervals will usually result in a graph that is not clear or well-defined. Example 5.3.1 demonstrates how the use of formats, interval multipliers, and shift operators can be used to make your graphs involving dates self-explanatory.

Example 5.3.1 Johnny's Savings Account

When Johnny turned 10 years old on September 1, 1975, he took all the money he had in his piggybank and deposited it into a bank account that paid 4.5% interest annually, compounded daily. He made a promise to add two dollars at the end of each week from any money that he earned, and that he would take the money out when he reached the ripe old age of 40. Johnny thought he might have $1,000 by then. He kept that promise, so let's look at Johnny's earnings from the time he was 10 until he was 40 using the following program:

```
TITLE "Johnny's 40th Birthday Fund";

PROC GPLOT DATA=book.graph1;
PLOT fund*date / VREF=5000 LV=1 CV=blue;
RUN;
```

Here is the resulting output:

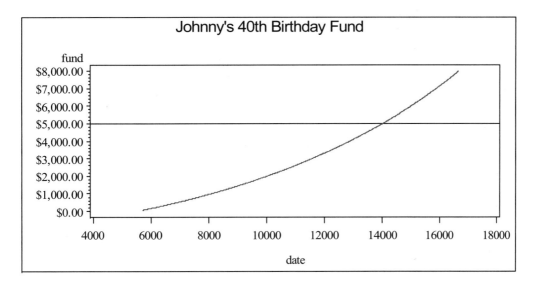

That's not very helpful. We can see that he started around 6000, and he turned 40 somewhere around 17000. What does that mean? This is easy enough to fix. Don't we just need to add a format to the dates like this?

```
TITLE "Johnny's 40th Birthday Fund";
PROC GPLOT DATA=book.graph1;
PLOT fund*date / VREF=5000 LV=1 CV=blue;
FORMAT date mmddyy10.; /* Add FORMAT statement */
RUN;
```

Here is the new output:

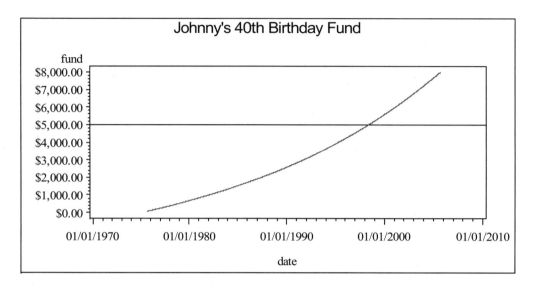

What has happened here is that SAS picked the boundaries and figured out major and minor tick marks. In this example, it has selected major intervals at decade boundaries, and minor ones at year boundaries. That's not a bad choice for this example, but SAS/GRAPH doesn't always make such a good choice when picking the boundaries of an axis. Can't you define the horizontal axis yourself?

Sure you can! Johnny's birth date is in September, so it would make more sense to chart his progress at his birthdays, and restrict the span of the horizontal axis to the period he's contributing. Let's shift the interval by nine months to fix the starting date and try this code:

```
TITLE "Johnny's 40th Birthday Fund";
PROC GPLOT DATA=book.graph1;
PLOT fund*date / VREF=5000 LV=1 CV=blue
               HAXIS='01SEP1975'd TO '01SEP2005'd by YEAR.9;
FORMAT date mmddyy10.;
RUN;
```

Here is the resulting output:

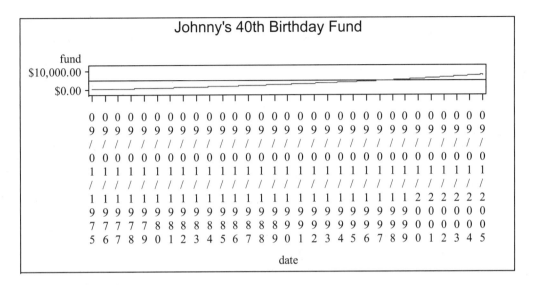

What happened here? Well, since we defined the horizontal axis as having tick marks *every* year, SAS accommodated our request. It was even thoughtful enough to turn the labels vertically to make sure that they all fit. Unfortunately, that left very little space for the graph itself, because SAS thinks the labels are more important than the graph. We need better spacing on our horizontal axis.

Since decades seemed to work well, let's use those as our intervals, but we want to start on Johnny's tenth birthday, and define the major horizontal axis points at September 1, 1975, September 1, 1985, September 1, 1995, and September 1, 2005. An interval multiplier of 10 will create the decade interval, and a shift operator of 69 (60 (months in 5 years) plus 9 (months from January to September)) will move the starting date of the ten-year interval to September 1975 so that it matches the starting point of the horizontal axis. Note that the only change from the previous version is in the interval definition.

```
TITLE "Johnny's 40th Birthday Fund";

PROC GPLOT DATA=book.graph1;
PLOT fund*date / VREF=5000 LV=1 CV=blue
     HAXIS='01SEP1975'd TO '01SEP2005'd BY YEAR10.69; /* Define X
                                                         axis with an
                                                         interval */

FORMAT date mmddyy10.;
RUN;
```

Here is the resulting output:

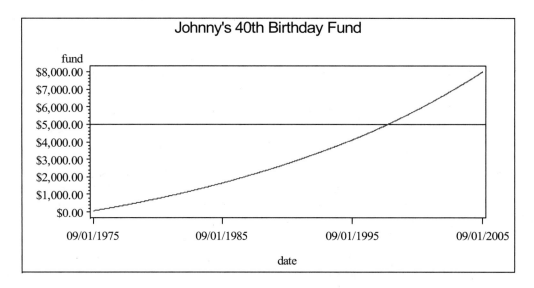

Now that's what we wanted. This demonstrates that you have all of the interval types, as well as their multipliers and shift operators, available when you are defining axes that involve date, time, and datetime values in SAS/GRAPH. It makes defining the exact scope of the graph much easier, not to mention comprehending what you've graphed.

5.4 The Basics of PROC EXPAND

The EXPAND procedure is a part of SAS/ETS, which is used with time series data. It creates a SAS data set, and does not routinely produce printed output. With Version 9 of SAS, you can use ODS for Statistical Graphics to produce output from PROC EXPAND, but it is an experimental product.

5.4.1 Capabilities of PROC EXPAND

It will change the sampling frequency of the data that you have and convert it to a different one. It can interpolate values in time series data, for example, when you have quarterly data that you need to report or analyze on a monthly basis. It can perform the reverse operation, that is, to aggregate (collapse) data from a higher sampling frequency to a lower one, such as taking monthly data and turning it into quarterly data. PROC EXPAND can interpolate missing values even if you aren't changing the sampling frequency. It also provides for extensive data transformations, and performs all of these functions without a lot of DATA step programming. The SAS/ETS documentation provides detail on the procedure, its statements, and the options for those statements.

PROC EXPAND uses SAS interval definitions. This includes shift operators and interval multipliers. For a detailed explanation of shift operators and interval multipliers, see Section 5.2. When you use these interval definitions (plus any shift and/or multipliers,) PROC EXPAND will automatically adjust for any calendar effects (leap years, varying number of days in a month). As with anything that uses these interval definitions, all measurements and calculations are considered to be at the beginning of the interval(s) specified. It is possible to change that definition with options in one of the PROC EXPAND statements, and those are discussed in Section 5.4.5.

Table 5.4.1 *PROC EXPAND Sample Data*

The following data set will be used for the examples in this section. This is light rail ridership data obtained from the American Public Transportation Association for the years 2003 and 2004, and is used with their permission. The values for October and November of 2003 have been removed to demonstrate some of the capabilities of PROC EXPAND.

date	Riders (thousands)
JAN2003	2679.9
FEB2003	2421.9
MAR2003	2704.6
APR2003	2778.3
MAY2003	2718.6
JUN2003	2618.2
JUL2003	2999.0
AUG2003	3504.7
SEP2003	3329.4
OCT2003	.
NOV2003	.
DEC2003	2888.6
JAN2004	3132.9
FEB2004	2814.3
MAR2004	3067.3
APR2004	2928.8
MAY2004	2958.3
JUN2004	2966.3
JUL2004	3000.8
AUG2004	3071.2
SEP2004	2958.9
OCT2004	2992.8
NOV2004	3017.5
DEC2004	3038.4

5.4.2 Using PROC EXPAND to Convert to a Higher Frequency

You can use PROC EXPAND to convert data from a lower sampling frequency to a higher sampling frequency (e.g., converting monthly data to daily or weekly data.) It does so by interpolation, and the syntax to convert to a higher sampling frequency is as follows:

```
1   PROC EXPAND DATA=book.month OUT=seven_days FROM=MONTH TO=WEEK;
2   ID date;
3   CONVERT riders;
4   RUN;
```

The PROC EXPAND statement in line 1 specifies the output data set ("seven_days"), and explains how the data in BOOK.MONTH should be converted, from MONTH intervals to WEEK intervals. The ID statement in line 2 indicates the variable that identifies the time of each record.

IMPORTANT

You will usually use an ID statement with PROC EXPAND; otherwise, SAS will create an ID variable for the input records, and it will use the starting point of January 1, 1960, which may not be what you want. The CONVERT statement identifies the variable(s) to convert. You may also rename the variable(s) being converted in the output data set like this: **CONVERT input-var=output-var;** Here are the first eight observations from the data set SEVEN_DAYS produced by the above code:

date	Riders (thousands)
29DEC2002	2770.23
05JAN2003	2573.57
12JAN2003	2449.76
19JAN2003	2393.52
26JAN2003	2391.24
02FEB2003	2429.29
09FEB2003	2494.03
16FEB2003	2571.86

5.4.3 Using PROC EXPAND to Convert to a Lower Frequency

You can convert data to a lower frequency with PROC EXPAND in two ways: first, you can use the same syntax as with converting to a higher sampling frequency, except that the TO= interval would be of a lower sampling frequency. When you convert your data this way, PROC EXPAND performs interpolation for missing values using a curve fitting method, and allows conversion between intervals that aren't nested. A nested interval is one that fits wholly inside of another interval (e.g., days nest within weeks, because there are exactly seven days in a week, but weeks do not nest within months, because most months have partial weeks). The following program will interpolate any missing values in our data:

```
PROC EXPAND DATA=book.month OUT=quarterly FROM=MONTH TO=QTR;
ID date;
CONVERT riders;
RUN;
```

Obs	date	Riders (thousands) After Interpolation	Original Values From BOOK.MONTH
1	01JAN2003	2679.90	2679.9
2	01APR2003	2778.30	2778.3
3	01JUL2003	2999.00	2999.0
4	01OCT2003	2993.28	.
5	01JAN2004	3132.90	3132.9
6	01APR2004	2928.80	2928.8
7	01JUL2004	3000.80	3000.8
8	01OCT2004	2992.80	2992.8

The resulting dataset QUARTERLY (above) has eight observations, four for each year, synchronized on the QTR interval boundaries. As you can see, the data that were missing from our original dataset (for October of 2003) are interpolated.

The second method allows you to perform simple aggregation (addition) without interpolation of missing values. The AGGREGATE method always produces an exact result without interpolation, and it requires that the intervals be nested. This program shows the result of a simple aggregation on our sample data:

```
PROC EXPAND DATA=book.month OUT=annual FROM=MONTH TO=YEAR;
ID date;
CONVERT riders / METHOD=AGGREGATE;
RUN;
```

date	Riders (thousands)	
2003	.	There are 2 missing observations for this year, which yields a missing result.
2004	35947.5	Total across all 12 months.

Example 5.4.1 The Importance of the ID Statement in PROC EXPAND

The following PROC EXPAND step has no ID statement. Let's run it so that we can see the assumptions that SAS makes in its absence. This illustrates why the ID statement is almost always used with this procedure:

```
PROC EXPAND DATA=book.month OUT=ANNUAL FROM=MONTH TO=YEAR;
CONVERT riders;
RUN;
```

date	Riders (thousands)
01JAN1960	2679.9
01JAN1961	3132.9

How did we wind up with data for 1960 and 1961 when we used data from 2003 and 2004? In the absence of an ID variable to indicate the dates, SAS will create ID values to label the input records, and it will start from its zero point, January 1, 1960. This is why you usually use an ID statement with PROC EXPAND.

5.4.4 Using PROC EXPAND to Interpolate Missing Values

PROC EXPAND can also be used to interpolate missing values without converting frequencies. There are two ways to do this; use the one that fits your situation. If you are interpolating missing values at specific points in time, leave off the FROM= and TO= options, but make sure that you use an ID statement to indicate the variable that contains the time points of the observed values. The time points do not have to be evenly spaced, nor do you need a record for each time point within the interval. PROC EXPAND will read the values supplied in the ID variable and figure out the interval to use. Remember that the data for the months of October and November are missing in our sample table. The following program demonstrates:

```
PROC EXPAND DATA=book.month OUT=nomiss;
ID date;
CONVERT riders;
RUN;
```

date	Riders (thousands)
01JAN2003	2679.90
01FEB2003	2421.90
01MAR2003	2704.60
01APR2003	2778.30
01MAY2003	2718.60
01JUN2003	2618.20
01JUL2003	2999.00
01AUG2003	3504.70
01SEP2003	3329.40
01OCT2003	2993.28
01NOV2003	2788.68
01DEC2003	2888.60

The second method interpolates missing values in a time series to a specific interval. Use this when you are interpolating to a different interval than the one given in the ID variable. It requires the FROM= option, but leave off the TO= option, as shown here:

```
PROC EXPAND DATA=book.month OUT=nomiss2 FROM=MONTH;
ID date;
CONVERT riders;
RUN;
```

By default, the interpolation is performed by fitting the points to a cubic spline curve. You can request other methods of interpolation with the METHOD= option on the CONVERT statement, and these are detailed in the SAS/ETS documentation. PROC EXPAND will ignore observations that have missing values for the ID variable, even if there are data points for the CONVERT variable(s). Table 5.4.2 is a summary of what PROC EXPAND does when there are missing values for the ID variable and/or CONVERT data points.

Table 5.4.2 *How PROC EXPAND Handles Interpolation of Missing Values in Input Data*

ID variable	Data	PROC EXPAND Will
Missing	*Missing*	Interpolate
Not Missing	*Missing*	Interpolate
Missing	Not Missing	Ignore

5.4.5 The OBSERVED= Option for the CONVERT Statement in PROC EXPAND

As with the other uses of SAS date, time, and datetime intervals, the default for PROC EXPAND is to consider the values as being from the beginning of the intervals provided in the FROM= and TO= options. This is not always the case with real-world data, and it can cause very different results, especially if the values are not measured at the beginning of the given interval(s), or they do not represent a single observed value for a specific point in time. You can control how the SAS intervals are used through with the OBSERVED option on the CONVERT statement. There are five different values for the OBSERVED= option, as shown in Table 5.4.3:

Table 5.4.3 *Values for the OBSERVED= option*

BEGINNING	Beginning of the period
MIDDLE	Middle of the period
END	End of the period
TOTAL	Totals for the period
AVERAGE	Averages across the period

Example 5.4.2 shows how the different values of the OBSERVED= option affect our sample data when we increase and decrease the sampling frequency. The program below shows the code to increase the sampling frequency. It also shows how to rename your output variables in the CONVERT statement by placing an equals sign (=) after the dataset variable and providing the new variable name afterwards.

Example 5.4.2 **Effect of Different Values for OBSERVED= Option on Increased Frequency**

```
/* Create weekly datasets from monthly data using different OBSERVED=
options */
PROC EXPAND DATA=book.month OUT=seven1 FROM=MONTH TO=WEEK;
ID date;
CONVERT riders=beginning / OBSERVED=BEGINNING /* Stores result in
variable named 'beginning' */;
RUN;

PROC EXPAND DATA=book.month OUT=seven2 FROM=MONTH TO=WEEK;
ID date;
CONVERT riders=middle / OBSERVED=MIDDLE;
RUN;

PROC EXPAND DATA=book.month OUT=seven3 FROM=MONTH TO=WEEK;
ID date;
CONVERT riders=end / OBSERVED=END;
RUN;

PROC EXPAND DATA=book.month OUT=seven4 FROM=MONTH TO=WEEK;
ID date;
CONVERT riders=total / OBSERVED=TOTAL;
RUN;
```

```
PROC EXPAND DATA=book.month OUT=seven5 FROM=MONTH TO=WEEK;
ID date;
CONVERT riders=average / OBSERVED=AVERAGE;
RUN;

/* Put all EXPAND datasets together for side-by-side display */
DATA compare_lo;
MERGE seven1 seven2 seven3 seven4 seven5;
BY date;
RUN;
```

The data set COMPARE_LO is shown here:

DATE	BEGINNING	MIDDLE	END	TOTAL	AVERAGE
29DEC2002	2770.19	.	.	580.484	3260.80
05JAN2003	2573.61	.	.	597.974	2897.25
12JAN2003	2449.83	2708.88	.	608.442	2638.87
19JAN2003	2393.59	2536.09	.	613.169	2472.89
26JAN2003	2391.28	2436.82	2652.65	613.550	2384.91
02FEB2003	2429.28	2398.83	2506.22	610.979	2360.52
09FEB2003	2493.99	2409.06	2428.26	606.851	2385.33
16FEB2003	2571.80	2454.40	2406.10	602.561	2444.93
23FEB2003	2649.08	2521.78	2427.05	599.504	2524.93

As you can see, the OBSERVED= option has a very large effect on the results that PROC EXPAND yields. The BEGINNING column is the default, and that is the interpolation calculated if the numbers are measured at the beginning of the month. MIDDLE and END do the calculation as if the numbers were measured at the middle and the end of the month, respectively. That is why the interpolated values are missing in those columns in the above chart. The numbers are not available until the beginning of the interval (the beginning of the week containing the middle and end, respectively, of the month).

If you're measuring totals (which we are, since this is mass transit ridership data), the values are radically different. TOTAL means that the number being interpolated is not representative of a single point in the TO= interval, but that it is obtained across the duration of the TO= interval. Therefore, the numbers in that column of the above table represent the number of riders per week. AVERAGE considers the numbers to be the TO= interval average.

In contrast to Example 5.4.2, Example 5.4.3 shows the effect of the OBSERVED= option on a lower sampling frequency. Remember, since our original data had missing values, some interpolation will take place.

Example 5.4.3 Effect of Different Values for OBSERVED= Option on Lowered Frequency

---

DATE	BEGINNING	MIDDLE	END	TOTAL	AVERAGE
2003	2679.9	2764.22	2888.6	34476.22	2861.76
2004	3132.9	2972.48	3038.4	35947.50	2996.92

BEGINNING, MIDDLE, and END don't give us a very good idea of yearly ridership, because they are considering the entire ridership as occurring on the beginning, middle, or end of the FROM= interval. TOTAL is the value for the entire year, and AVERAGE is calculated on the monthly values. However, all of these values are calculated with interpolation of the missing values in October and November of 2003.

PROC EXPAND has many more capabilities, and the preceding examples give only the most basic information on how to use this powerful procedure with time series data. You can refer to the documentation for SAS/ETS to get a much more complete explanation of PROC EXPAND and its options.

5.5 International Date, Time, and Datetime Formats and Informats

Version 9 of SAS has formats for dates and times in languages other than United States English. It is included in Base SAS as a part of National Language Support (NLS). The key to NLS is in the LOCALE= or DFLANG= system options. The LOCALE= option is defined in the SAS configuration file when it is installed by your SAS administrator, but it may be changed with an OPTIONS statement, or inside the OPTIONS window. The LOCALE= option implicitly sets two other options which can affect dates, times, and datetime values in SAS. The DATESTYLE= option determines how the ANYDT informats will interpret character strings where month, day, and year are ambiguous. The DFLANG= option defines the default language that the SAS System will use.

There are a few specific date formats for Taiwanese, Japanese, and Hebrew, but you can consider the majority of international formats and informats as falling into one of two informal categories: the "EUR" category, or the "NL" category. These categories are based on the first two or three letters of the format or informat name. The "EUR" category will select the language based on either the DFLANG= system option or by allowing you to replace the "EUR" in the format name with a specific language abbreviation. Using the language abbreviations is handy if you are working with many languages on the same output, because they allow you to specify the language without regard to the DFLANG= option. The "NL" category is controlled by the LOCALE= system option.

5.5.1 "EUR" Formats and Informats

Each of these formats and informats correspond to an English language format or informat. However, the minimum, maximum, and default widths for the format or informat are dependent upon the language being used at the time. Tables 5.5.1 and 5.5.2 list the English language formats and their EUR format names, and the EUR informats.

Table 5.5.1 *International Format Names and Their English Language Equivalents*

English language format name	International format name
DATE.	EURDFDE.
DATETIME.	EURDFDT.
DDMMYY.	EURDFDD.
DOWNAME.	EURDFDWN.
MONNAME.	EURDFMN.
MONYY.	EURDFMY.
WEEKDATX.	EURDFWKX.
WEEKDAY.	EURDFDN.
WORDDATX.	EURDFWDX.

Table 5.5.2 *International Informat Names and Their English Language Equivalents*

EURDFDE**w**.	Reads international date values in the form ddmonyy(yy), where dd represents the day of the month, mon is the three-letter month abbreviation in the language specified by the DFLANG= system option, or by the appropriate three-letter prefix, and yy(yy) is the two- or four-digit year.
EURDFDT**w**.	Reads international datetime values in the form ddmonyy hh:mm:ss.ss or ddmonyyyy hh:mm:ss.ss
EURDFMY**w**.	Reads month and year date values in the form monyy or monyyyy

You may replace the "EUR" with a specific three-letter language prefix in any of the above formats or informats to define the language that you wish to use. This overrides the DFLANG= system option, and is a good way to display dates in multiple languages simultaneously. Table 5.5.3 is a list of all the valid languages with their three-letter prefix. In addition, we'll show the effect of using each three-letter prefix on the EURDFWKX. format by using the reference date of Monday, August 18, 1997. As a comparison, Table 5.5.3 also includes the reference date formatted as the English equivalent WEEKDATX.

Table 5.5.3 *International Date Formats with Language Abbreviations*

LANGUAGE PREFIX	LANGUAGE	FORMAT NAME	FORMATTED DATE
		WEEKDATX.	Monday, 18 August 1997
AFR	Afrikaans	AFRDFWKX.	Maandag, 18 Augustus 1997
CAT	Catalan	CATDFWKX.	Dilluns, 18 Agost 1997
CRO	Croatian	CRODFWKX.	ponedjeljak, 18 kol 1997
CSY	Czech	CSYDFWKX.	pondělí, 18 srpen 1997

(continued on next page)

Table 5.5.3 (continued)

LANGUAGE PREFIX	LANGUAGE	FORMAT NAME	FORMATTED DATE
DAN	Danish	DANDFWKX.	mandag, den 18. august 1997
NLD	Dutch	NLDDFWKX.	maandag, 18 augustus 1997
FIN	Finnish	FINDFWKX.	Maanantaina, 18. elokuuta 1997
FRA	French	FRADFWKX.	Lundi 18 août 1997
DEU	German	DEUDFWKX.	Montag, 18. August 1997
HUN	Hungarian	HUNDFWKX.	1997.augusztus 18., hétfő
ITA	Italian	ITADFWKX.	Lunedì, 18 Agosto 1997
MAC	Macedonian	MACDFWKX.	ponedelnik, 18 avgust 1997
NOR	Norwegian	NORDFWKX.	mandag, 18 august 1997
POL	Polish	POLDFWKX.	poniedziałek, 18 sierpień 1997
PTG	Portuguese	PTGDFWKX.	Segunda-feira, 18 de agosto de 199
RUS	Russian	RUSDFWKX.	Понедельник, 18 Август 1997
ESP	Spanish	ESPDFWKX.	lunes, 18 de agosto de 1997

(*continued on next page*)

Table 5.5.3 (continued)

LANGUAGE PREFIX	LANGUAGE	FORMAT NAME	FORMATTED DATE
SLO	Slovenian	SLODFWKX.	ponedeljek, 18 avgust 1997
SVE	Swedish	SVEDFWKX.	Måndag, 18 augusti 1997
FRS	Swiss_French	FRSDFWKX.	Lundi 18 août 1997
DES	Swiss_German	DESDFWKX.	Montag, 18. August 1997

5.5.2 "NL" Formats

The output from the "NL" series of formats is defined by the LOCALE= system option. Unlike the "EUR" series, you cannot specify a language other than the one defined by the current value of the LOCALE= option. Use these formats when your output may be generated in several different locations around the world, but you don't have to display multiple languages within the same output.

These formats work by converting the SAS date, time, or datetime value to that of the specified locale, and then formatting the result. These formats are also noteworthy in that the result is *left-justified*, as opposed to the right-justification of most of the other date, time, and datetime formats. This is true for all ODS destinations as well as for traditional column-based output. Of course, with ODS destinations, the justification of the column will be performed according to any STYLE in effect. Table 5.5.4 lists the "NL" series formats available and the default width, width range, and English language equivalent for each format.

Table 5.5.4 *"NL" Series Formats*

CATEGORY	FORMAT NAME	ENGLISH LANGUAGE FORMAT NAME	DEFAULT WIDTH	WIDTH RANGE	
Date	NLDATE**w**.	DATE.	20	20–100	
	NLDATEMN**w**.	MONNAME.	4	4–200	
	NLDATEW**w**.	WORDDATE. or WORDDATX.	20	20–200	
	NLDATEWN**w**.	DOWNAME.	10	4–200	
Datetime	NLDATM**w**.	DATETIME.	30	10–200	**!** IMPORTANT
	NLDATMAP**w**.	DATEAMPM.	32	16–200	
	NLDATMTM**w**.	**None.** Displays time-of-day from datetime value in local time format.	16	16–200	Unlike the English-language formats, fractional seconds may not be used.
	NLDATMW**w**.	DTWKDATX.	30	16–200	
Time	NLTIMAP**w**.	TIMEAMPM.	10	4–200	
	NLTIME**w**.	TIME.	20	10–200	

Table 5.5.5 shows the difference between two different LOCALE settings when each of the following formats are used.

Table 5.5.5 *Differences in the "NL" Formats Based on LOCALE= Option Setting*

FORMAT NAME	OPTIONS LOCALE= ENGLISH_US	OPTIONS LOCALE= SPANISH_SPAIN
NLDATE32.	April 16, 2006	16 de abril de 2006
NLDATEMN32.	April	abril
NLDATEW32.	Sunday, April 16, 2006	domingo 16 de abril de 2006
NLDATEWN32.	Sunday	domingo
NLDATM32.	16Apr04:18:25:45	16 de abril de 2004 18H25
NLDATMAP32.	April 16, 2004 06:25:45 PM	16 de abril de 2004 18H25
NLDATMTM32.	18:25:45	18H25
NLDATMW32.	Fri, Apr 16, 2004 06:25:45 PM	vie 16 de abr de 2004 18H25
NLTIMAP32.	06:25:45 PM	18H25
NLTIME32.	18:25:45	18H25

Table 5.5.6 shows the available "NL" series informats, their default width specification, the width range, and the English language informat to which it is similar.

Table 5.5.6 *"NL" Series Informats*

CATETORY	FORMAT NAME	DEFAULT WIDTH	WIDTH RANGE	ENGLISH LANGUAGE EQUIVALENT INFORMATS
Date	NLDATE*w*.	20	20–100	DATE. and WORDDATX.
Datetime	NLDATM*w*.	30	10–200	DATETIME.
Time	NLTIMAP*w*.	10	4–200	TIME., with AM/PM
	NLTIME*w*.	20	10–200	TIME.

SUMMARY

The whole point of the "NL" series of formats and informats is that you do not have to worry about the specific language that will be using the format or informat. The LOCALE= option will take care of it. In this way, the same SAS program can be used anywhere, and the output will be appropriate to the local language. It is important to understand that the "NL" series of formats and informats works just as well with the English language, so there's no need to use SAS program code beyond the LOCALE= option to switch between English language formats/informats and other languages, unless you need a specific English language format that does not have an "NL" equivalent.

5.5.3 *Specific Language Date Formats and Informats*

In addition to the "NL" and "EUR" series of formats and informats, SAS has a few other date formats and informats for specific languages. Some of these are available in Version 8, and are noted as such.

5.5.3.1 Hebrew Date Formats (Version 9)

HDATE*w.* displays a SAS date value in Hebrew. You will need the correct character encoding installed on your system to display this correctly. The SAS date will be displayed as **yyyy mmmmm dd**, where **dd** is the day-of-the-month, **mmmmm** represents the month's name in Hebrew, and **yyyy** is the year. **w** can be from 9–17, with a default width of 17, and it is right-justified. Use odd numbers for **w** to get the best display.

HEBDATE*w.* displays a SAS date value according to the Jewish calendar. **w** can be from 7–24, with a default width of 16, and it is right-justified. There are three forms of the display, long, default, and short, dependent upon the width specified. Again, you will need the correct character encoding installed on your system, or you will get substitutions for non-printing characters.

5.5.3.2 Japanese and Taiwanese Date Formats (Versions 8 and above)

MINGUO*w.* displays a SAS date value as a Taiwanese date value in the form **yy(yy)mmdd**, where **yy(yy)** is the year, **mm** is the number of the month, and **dd** is the day of the month. **w** can range from 1–10, with a default width of 8, and it is **left-justified**, and zero-filled. The Taiwanese calendar uses 1912 as the base year (i.e., 01/01/01 is January 1, 1912). Also, the year values continue to increase past 100; they do not cycle. January 1, 2012 is "100/01/01", not "00/01/01".

NENGOw. writes a SAS date value in the form **e.yymmdd**, where **e** is the first letter of the name of the emperor (Meiji, Taisho, Showa, or Heisei), **yy** is the year, **mm** is the month, and **dd** is the day of the month. **w** can be from 2–10, with a default width of 10, and it is **left-justified**. SAS will omit the period if **w** isn't big enough.

5.5.3.3 Japanese and Taiwanese Date Informats (Versions 8 and above)

JDATEMYDw. allows you to convert Japanese Kanji in the form yy(yy)mondd to SAS date values, where **yy(yy)** is the year, **mon** is the Kanji representation of the name of the month, and **dd** represents the day of the month. **w** can be from 12–32, with a default width of 12. You can separate (yy)yy, mon, and dd with special characters or blanks, but you must make sure that the width specification allows for any blanks and/or special characters in the input field. Two-digit years will be translated according to the YEARCUTOFF= option.

JNENGOw. reads Japanese Kanji date values in the form yymmdd, where **yy** is the year, **mm** is the Kanji representation of the name of the month, and **dd** represents the day of the month. Since yy is two digits long, this informat is always affected by the YEARCUTOFF= option. **w** can be from 16–32, with a default width of 16. You can separate **yy**, **mon**, and **dd** with special characters or blanks, but you must make sure that the width specification allows for any blanks and/or special characters in the input field.

MINGUOw. converts a Taiwanese date value into a SAS date value in the form **yy(yy)mmdd**, where **yy(yy)** is the year, **mm** is the number of the month, and **dd** is the day of the month. **w** can be from 6–10, with a default width of 6. You may use separators such as blanks, dashes, or slashes between the year, month, and day values, but they must be present between all of the values. The Taiwanese calendar uses 1912 as the base year (i.e., 01/01/01 is January 1, 1912). In addition, the year values continue to increase past 100; they do not cycle. January 1, 2012 is "100/01/01", not "00/01/01".

5.6 Other Software and Their Dates (Excel, Oracle, DB2)

Most other software packages keep their dates in some sort of numerical form in much the same way that SAS does, while some software packages have a special variable type for dates. Microsoft Excel stores dates as integers, but it uses January 1, 1900 instead of January 1, 1960 as day zero. Times are stored in Excel as fractions of days, so noon of a given day is .5 (exactly one half of a day.) Datetime values are stored in Excel as the day relative to 01/01/1900 plus the fraction of the day. In Excel, 6 p.m., on January 1, 1900 is represented as .75. Excel also has a major limitation on its dates: it cannot store dates as

CAUTION

negative numbers, so any date prior to January 1, 1900 is going to be represented by a character string, not an Excel date value. This may cause problems if you are importing data from Excel. If you want to do the conversion from/to SAS date, time, and datetime values to/from Excel date and time values, you can use the following conversion table:

Table 5.6.1 *Mathematical Conversions between SAS and Excel*

To convert Excel values into SAS values
Creating a SAS date: Subtract 21916 from the Excel date value.
Creating a SAS time: Multiply the Excel time value (fraction of a day) by 86400 (# of seconds in a day).
Creating a SAS datetime: Subtract 21916 from the Excel date and time value, then multiply by 86400;
To convert SAS values into Excel values
Creating an Excel date value: Add 21916 to the SAS date value. **This works only for dates after December 31, 1899.**
Creating an Excel time value: Divide the SAS time value by 86400.
Creating an Excel date and time value: Divide the SAS datetime value by 86400 (seconds in a day), then add 21916 to that. Again, this works only for dates after December 31, 1899.

What do you do if you have dates before January 1, 1900 to convert? Going to Excel, you will have to store them as character strings, and you won't be able to use any of the Excel math functions on them. On the other hand, if you are going from Excel to SAS, then you can take the character string from the imported data set and use the INPUT() function (Section 3.3.2) to get a valid SAS date or datetime value. The ANYDT informats may also prove useful in situations like this. Example 5.6.1 shows how to use an INPUT statement and a DATA step to process a CSV file when you have mixed character and date values from Excel. Note that we have to use the OPTIONS YEARCUTOFF= statement to process the two-digit year values in the sample file correctly.

Example 5.6.1 Reading an Excel CSV File with Dates Prior to January 1, 1900

THE CSV FILE "EXCEL_TEST.CSV"
01/01/00
01JAN1899
01/01/01

Not an Excel date value, so it is not in the Excel date format for the column.

The Log

```
117   OPTIONS YEARCUTOFF=1895; /* If not specified, the first and last dates
in the "excel_test.csv" file will be in the 21ˢᵗ century! */




118   OPTIONS DATESTYLE=DMY;
119   DATA convert_excel;
120   INFILE "examples\excel_test.csv" PAD MISSOVER;
121   INPUT @1 date anydtdte10.;
122   PUT _infile_;
123   PUT date= +5 date= worddate.;
124   RUN;

NOTE: The infile "examples\excel_test.csv" is:
      File Name=C:\book\examples\excel_test.csv,
      RECFM=V,LRECL=256

01/01/00
date=-21914        date=January 1, 1900

01JAN1899
date=-22279        date=January 1, 1899

01/01/01
date=-21549        date=January 1, 1901
```

Since the dates are in the CSV file with two different formats (DATE9. and MMDDYY.), we can't use either of those informats to process the input field. The ANYDTDTE. informat allows us to process this easily.

These are conversion issues specific to Excel that may arise when you are trying to import or export data to/from Excel. When you import data from other packages into SAS using the IMPORT procedure, or one of the database engines, SAS should understand and convert the dates, even though the reference date may differ. There are exceptions to this rule, one of which is using a WHERE clause inside PROC SQL for foreign databases. You will have to know the date, time, or datetime format for the foreign database to select records based on dates, times, or datetimes.

However, before assuming that your dates are numeric (or of type "date"), make sure that you are not working with character strings masquerading as dates. If you have a character string, you will have to convert it to a SAS date, time, or datetime yourself with the INPUT() function (Section 3.3.2)

Sending dates to other databases and software packages should be fine if you use the EXPORT procedure or one of the database engines. If you are determined to send dates to another database or software package the hard way, then you will have to produce SAS date, time, or datetime values as character strings in the format of the other software. You can use a picture format and the PUT statement to accomplish this, as long as you know the correct representation of the package for which you are creating the data. For details on creating picture formats, see Sections 2.5 and 2.6. Example 5.6.2 shows how this is done for a DB2 database.

Example 5.6.2 Writing Datetime Values for DB2 Using a Picture Format

```
PROC FORMAT;
PICTURE dbdate
LOW-HIGH = '%Y-%m-%d:%H:%M:%S' (DATATYPE=DATETIME)
. - .Z = '0000-00-00:00:00:00';
RUN;

DATA _NULL_;
now = '01JUL2005:20:18:32'dt;
PUT "now displayed as datetime value: " now;
PUT "now displayed as datetime18.: " now DATETIME18.;
PUT "now displayed as dbdate.: " now DBDATE.;
RUN;
```

The Log

```
now displayed as datetime value: 1435868312
now displayed as datetime18.:   01JUL05:20:18:32
now displayed as dbdate.:    2005-07-01:20:18:32
```

APPENDIX A A Quick Reference Guide to SAS Date, Time, and Datetime Formats

This table shows the result when the same date, time, or datetime value is displayed with the corresponding format, using the default length for the given format.

The reference date for this table is Sunday, June 27, 2004.

If you want your date to look like this	Use this format	If you want your date to look like this	Use this format
27JUN04	DATE.	1	WEEKDAY.
27	DAY.	2004-W26-01	WEEKU.
27/06/04	DDMMYY.	2004-W26-07	WEEKV.
27 06 04	DDMMYYB.	2004-W25-07	WEEKW.
27:06:04	DDMMYYC.	June 27, 2004	WORDDATE.
27-06-04	DDMMYYD.	27 June 2004	WORDDATX.
27062004	DDMMYYN.	2004	YEAR.
27.06.04	DDMMYYP.	2004:06	YYMMC.
Sunday	DOWNAME.	2004-06	YYMMD.
179	JULDAY.	200406	YYMMN.
04179	JULIAN.	2004.06	YYMMP.
06/27/04	MMDDYY.	04-06-27	YYMMDD.
06 27 04	MMDDYYB.	04 06 27	YYMMDDB.
06:27:04	MMDDYYC.	04:06:27	YYMMDDC.
06-27-04	MMDDYYD.	04-06-27	YYMMDDD.
06272004	MMDDYYN.	20040627	YYMMDDN.
06.27.04	MMDDYYP.	04.06.27	YYMMDDP.
06M2004	MMYY.	2004JUN	YYMON.
06:2004	MMYYC.	2004Q2	YYQ.
06-2004	MMYYD.	2004:2	YYQC.
062004	MMYYN.	2004-2	YYQD.
06.2004	MMYYP.	20042	YYQN.
June	MONNAME.	2004.2	YYQP.
6	MONTH.	2004QII	YYQR.
JUN04	MONYY.	2004:II	YYQRC.
2	QTR.	2004-II	YYQRD.
II	QTRR.	2004II	YYQRN.
Sunday, June 27, 2004	WEEKDATE.	2004.II	YYQRP.
Sunday, 27 June 2004	WEEKDATX.		

The reference time for this table is 2:45 PM.

If you want your time to look like this	Use this format
14:45	HHMM.
15	HOUR.
885	MMSS.
14:45:00	TIME.
2:45:00 PM	TIMEAMPM.
14:45:00	TOD.

The reference datetime for this table is 2:45 PM on Sunday, June 27, 2004.

If you want your time to look like this	Use this format
27JUN04:02:45:00 PM	DATEAMPM.
27JUN04:14:45:00	DATETIME.
27JUN04	DTDATE.
JUN04	DTMONYY.
Sunday, 27 June 2004	DTWKDATX.
2004	DTYEAR.
04:2	DTYYQC.

Index

Books Available from SAS® Press

support.sas.com/pubs

*A Handbook of Statistical Analyses Using SAS®,
Second Edition*
by **B.S. Everitt**
and **G. Der**

Health Care Data and SAS®
by **Marge Scerbo, Craig Dickstein,**
and **Alan Wilson**

The How-To Book for SAS/GRAPH® Software
by **Thomas Miron**

*In the Know... SAS® Tips and Techniques From Around
the Globe*
by **Phil Mason**

*Instant ODS: Style Templates for the Output
Delivery System*
by **Bernadette Johnson**

*Integrating Results through Meta-Analytic Review Using
SAS® Software*
by **Morgan C. Wang**
and **Brad J. Bushman**

Learning SAS® in the Computer Lab, Second Edition
by **Rebecca J. Elliott**

The Little SAS® Book: A Primer
by **Lora D. Delwiche**
and **Susan J. Slaughter**

The Little SAS® Book: A Primer, Second Edition
by **Lora D. Delwiche**
and **Susan J. Slaughter**
(updated to include Version 7 features)

The Little SAS® Book: A Primer, Third Edition
by **Lora D. Delwiche**
and **Susan J. Slaughter**
(updated to include SAS 9.1 features)

The Little SAS® Book for Enterprise Guide 3.0
by **Susan J. Slaughter**
and **Lora D. Delwiche**

*Logistic Regression Using the SAS® System:
Theory and Application*
by **Paul D. Allison**

Longitudinal Data and SAS®: A Programmer's Guide
by **Ron Cody**

Maps Made Easy Using SAS®
by **Mike Zdeb**

Models for Discrete Date
by **Daniel Zelterman**

*Multiple Comparisons and Multiple Tests Using
SAS® Text and Workbook Set*
(books in this set also sold separately)
by **Peter H. Westfall, Randall D. Tobias,
Dror Rom, Russell D. Wolfinger,**
and **Yosef Hochberg**

Multiple-Plot Displays: Simplified with Macros
by **Perry Watts**

*Multivariate Data Reduction and Discrimination with
SAS® Software*
by **Ravindra Khattree**
and **Dayanand N. Naik**

Output Delivery System: The Basics
by **Lauren E. Haworth**

*Painless Windows: A Handbook for SAS® Users,
Third Edition*
by **Jodie Gilmore**
(updated to include Version 8 and SAS 9.1 features)

The Power of PROC FORMAT
by **Jonas V. Bilenas**

PROC SQL: Beyond the Basics Using SAS®
by **Kirk Paul Lafler**

PROC TABULATE by Example
by **Lauren E. Haworth**

*Professional SAS® Programmer's Pocket Reference,
Fifth Edition*
by **Rick Aster**

*Professional SAS® Programming Shortcuts,
Second Edition*
by **Rick Aster**

support.sas.com/pubs

Quick Results with SAS/GRAPH® Software
by **Arthur L. Carpenter**
and **Charles E. Shipp**

Quick Results with the Output Delivery System
by **Sunil Gupta**

Reading External Data Files Using SAS®: Examples Handbook
by **Michele M. Burlew**

Regression and ANOVA: An Integrated Approach Using SAS® Software
by **Keith E. Muller**
and **Bethel A. Fetterman**

SAS® for Forecasting Time Series, Second Edition
by **John C. Brocklebank**
and **David A. Dickey**

SAS® for Linear Models, Fourth Edition
by **Ramon C. Littell, Walter W. Stroup,**
and **Rudolf Freund**

SAS® for Mixed Models, Second Edition
by **Ramon C. Littell, George A. Milliken, Walter W. Stroup,** and **Russell D. Wolfinger**

SAS® for Monte Carlo Studies: A Guide for Quantitative Researchers
by **Xitao Fan, Ákos Felsővályi, Stephen A. Sivo,**
and **Sean C. Keenan**

SAS® Functions by Example
by **Ron Cody**

SAS® Guide to Report Writing, Second Edition
by **Michele M. Burlew**

SAS® Macro Programming Made Easy
by **Michele M. Burlew**

SAS® Programming by Example
by **Ron Cody**
and **Ray Pass**

SAS® Programming for Researchers and Social Scientists, Second Edition
by **Paul E. Spector**

SAS® Programming in the Pharmaceutical Industry
by **Jack Shostak**

SAS® Survival Analysis Techniques for Medical Research, Second Edition
by **Alan B. Cantor**

SAS® System for Elementary Statistical Analysis, Second Edition
by **Sandra D. Schlotzhauer**
and **Ramon C. Littell**

SAS® System for Regression, Third Edition
by **Rudolf J. Freund**
and **Ramon C. Littell**

SAS® System for Statistical Graphics, First Edition
by **Michael Friendly**

The SAS® Workbook and *Solutions* Set
(books in this set also sold separately)
by **Ron Cody**

Selecting Statistical Techniques for Social Science Data: A Guide for SAS® Users
by **Frank M. Andrews, Laura Klem, Patrick M. O'Malley, Willard L. Rodgers, Kathleen B. Welch,**
and **Terrence N. Davidson**

Statistical Quality Control Using the SAS® System
by **Dennis W. King**

A Step-by-Step Approach to Using the SAS® System for Factor Analysis and Structural Equation Modeling
by **Larry Hatcher**

A Step-by-Step Approach to Using SAS® for Univariate and Multivariate Statistics, Second Edition
by **Norm O'Rourke, Larry Hatcher,**
and **Edward J. Stepanski**

Step-by-Step Basic Statistics Using SAS®: Student Guide and *Exercises*
(books in this set also sold separately)
by **Larry Hatcher**

support.sas.com/pubs

Survival Analysis Using SAS®:
A Practical Guide
by **Paul D. Allison**

Tuning SAS® Applications in the OS/390 and z/OS
Environments, Second Edition
by **Michael A. Raithel**

Univariate and Multivariate General Linear Models:
Theory and Applications Using SAS® Software
by **Neil H. Timm**
and **Tammy A. Mieczkowski**

Using SAS® in Financial Research
by **Ekkehart Boehmer, John Paul Broussard,**
and **Juha-Pekka Kallunki**

Using the SAS® Windowing Environment:
A Quick Tutorial
by **Larry Hatcher**

Visualizing Categorical Data
by **Michael Friendly**

Web Development with SAS® by Example
by **Frederick Pratter**

Your Guide to Survey Research Using the
SAS® System
by **Archer Gravely**

JMP® Books

JMP® for Basic Univariate and Multivariate Statistics:
A Step-by-Step Guide
by **Ann Lehman, Norm O'Rourke, Larry Hatcher,**
and **Edward J. Stepanski**

JMP® Start Statistics, Third Edition
by **John Sall, Ann Lehman,**
and **Lee Creighton**

Regression Using JMP®
by **Rudolf J. Freund, Ramon C. Littell,**
and **Lee Creighton**

support.sas.com/pubs

Titles in Art Carpenter's SAS® Software Series

**Quick Results with the
Output Delivery System
by Sunil K. Gupta**
(Order No. A58458)

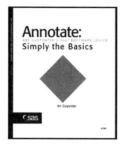

**Annotate: Simply the Basics
by Art Carpenter**
(Order No. A57320)

**Multiple-Plot Displays:
Simplified with Macros
by Perry Watts**
(Order No. A58314)

**Maps Made Easy
Using SAS®
by Mike Zdeb**
(Order No. A57495)

**The Power of
PROC FORMAT
by Jonas V. Bilenas**
(Order No. A59498)

To order: support.sas.com/pubs
Or call: (800) 727-3228